I0471641

# IMPLEMENT INFINITE IMAGINATION; CREATIVELY CONSTRUCT COSMOS

Ashraf A.

Ashraf A.

*A brightly brilliant book of seriously significant scientific literature...*

*"Invent Infinite Imagination of Marvellously Magnificent Mind*
*by Understanding Ultimately Unlimited Universe to*
*Creatively Construct Complete Cosmos Inside Infinite Imagination*
*by Indefinitely Implementing Infinite Imagination of Yours."*

# INITIAL INTRODUCTION

The human brain is an extremely powerful infinite quantum computer that can be programmed with human language. First, let me program my own brain; so that I can complete this book in noble nineteen nights. I already completed a book named "Admirably Amicable Assonance", a unique book of alliteration that can increase imagination; intelligence; creativity by an enormous amount. Completing that book immensely increased innermost intelligence of my mind; so whenever I decide to do something serious; I just do it and I do not quit until I complete it; quitting is giving up to accept defeat; long lasting loser always accept disgracefully dreadful defeat; so, I changed my mind's meaning of "give up" from "let someone, something, specially supporting spirits of the heart to fully fall down, downward toward the darkness dumping districts; depositing darkish desires; dark depressions; darkish delusions" to "gloriously give or lively lift superbly supreme supporting spirits up, upward ultimately unto unlimited unbounded Universes. I decided that I will write this book; a book that will explain the scientific knowledge of Cosmos; I will explain the wonderful science of the cosmos so simply that any human with basic understanding of English language will be able to understand it. I will complete this book no matter what happens; no matter what others say; no matter what difficulty comes.

No solution can exist without problems; it is the problem that inspires the solutions; then finally, those solutions give meaning to every problem. Problems and difficulties can never defeat us; neither they can control us but we are noble human beings, who can control us; our problems to solve them; to achieve triumphant win; every ethically moral human is a noble human being. No matter what difficulty you are facing in your life remember that you are the only one, who can program your own mind to solve it; you just have to try; so try harder. I want to say few simple words to those humans who think that their problems have no solution because every human mind is valuable.

"You are a highly honorable human with a brightly brilliant brain; who can solve any puzzling problem; discover any significant solution; achieve any accomplishment; acquire any authentic knowledge; understand any important information; because you are a heavenly human; a cosmic creationist containing a marvelously magnificent mind; an extremely powerful mind that can create constantly continuous; celestially captivating; colossal cosmos inside its indefinitely immense infinite imagination by intensely implementing its invaluable imaginative intelligence."

Am I programming your brain? Yes, I am but you will have to accept it too; because this is the specialty of a human brain that it needs to accept the prestigiously perfect programming and apply it to itself; into itself; inside itself; brilliantly by itself.

Why understanding Unlimited Universe is important?

The answer is very simple because its relative; relativity everywhere; not only in physics but everywhere, trust me. The man who discovered relativity was a very wise man who used "Relativity"; a cosmic word; Universal truth everywhere infinitely; even infinity is relative. Understanding unboundedly unlimited Universe gives any human the capabilities to understand self-value; forever fortunate freedom's value; humanity's value; heavenly home planet's value relative to Colossal Cosmos.

"The insanity is infinite and the sanity is the region enclosed by a small sphere within that dark infinite insanity; any area within any finite sphere of sanity is small relative to infinite insanity; we morally sane humans must always tremendously try to keep ourselves within that small sphere of stable sanity, otherwise we will lose ourselves inside infinite insanity; Only the eternal existence of love, peace, freedom and wisdom can keep us within the small sphere of essentially ethical stable sanity."

The cosmos is infinite relative to humanity. I will be using "Relativity" and "Relative" words in many places of this book. Am I confusing you? No, I am using relativity on your brain to program it to understand the cosmos. This planet Earth is the only home of humanity; our only cosmic address; our only outstandingly amicable accommodation. This planet is very small relative to the universe and the universe is infinite relative to our home planet.

Your own home in your city in your country is your address. Your country is your birthplace and this planet is the birthplace of your country; the Earth is the birthplace of humanity and you are an important infant of honorable humanity; the Universe is the birthplace of the Earth; the Cosmos is the birthplace of your authentic actuality; remarkable reality; essential entirety.

If you wake up one day in an unknown place and also find out that you have no knowledge of your address; your home; your city; your country; your birthplace, then your own mind will perfectly program insanity on itself. So it's better for you that you never forget that important information. The Earth and the Universe is also your cosmic address so you must also have some knowledge of it. The Cosmos is the identity of your authentically acknowledged actuality; remarkably rightful reality; essentially eminent entirety; so you must also try to understand it. You can never discover yourself without knowing your entire identity.

Relativistic knowledge of cosmos can make you understand that we humans are living in a small planet called Earth; so we must live here peacefully; we must protect ourselves and the entire humanity and all arriving and living lives; we must protect the habitability of our home planet Earth. So, always ask your youthfully curious Cosmic Consciousness cleverly, can you refuse your actual entirety? Can you refuse to implement infinite imagination that can indefinitely increase intelligence; the power of your marvelously magnificent mind? Can you refuse to understand the true home and its value?

Human's cosmic consciousness cannot cravingly collect perfect peace elsewhere except Earth's extremely essential; enormously eminent; excellently extraordinary environment; encouraging endless evolution; ensuring everlasting existence of honorable human's humanity. So, humans must always perfectly protect the Erath's excellently essential environment and appropriately amicable atmosphere.

May you let me program your powerful brain furthermore? Trust me, that human brain of yours is very powerful; Even powerful than an advance quantum computer. There are humans in this timeline who already understand the infinite cosmos; they are humans like you with a human brain. Only infinity can completely sustain infinity; only infinite imagination can completely imagine infinity and an imaginatively inventive; intelligently infinite cosmic consciousness can understand the knowledge of infinity. The Human Mind is intelligently infinite because it can understand infinity; so you can capably understand the infinity too; soon your mind will become infinite if you learn to infinitely increase intelligence to implement infinite imagination to completely create constantly continuous Cosmos. Remember, all cosmic objects are variables relative to human understanding. Cosmos existed before us; we humans arrived here after a very long period of time. To understand, the mysterious objects of the cosmos our human minds accepted those objects as variables and then relative to our own knowledge of understanding, we assigned values to those variables. Variables are just names and the values are the particular properties. If you take the Earth as an object than the name "EARTH" is just a variable like every other thing in the cosmos.

One of the properties of the Earth is that it's a planet that orbiting a star named the Sun. The most important thing is that you must prepare your mind or in other words, you must program your brain by telling it that you will go beyond by truly trying and understanding because human brain exists to understand; then it will accept infinite knowledge of the cosmos or anything you need most. My goal is to give you the capability to understand the power of infinite imagination so that you can create complete cosmos inside infinite imagination.

"Brilliantly breaking boundaries of unlimited unknown uncertainties by understanding unsurpassable Unlimited Universe while wonderfully surviving surely is significantly supreme success of honorable humans, heavenly Humanity.

# SpaceTime's Simple Structure

We will begin with Space-Time. Space and time are not
separated things but combined together. Space cannot exist
without time and time cannot exist without space. When an
object moves in Space it also moves in Time. There are three
spatial dimensions; you can imagine them as X, Y, Z. The X and
Y dimensions together imagined as a plane, like a plain paper
and then the Z dimension is the height. Together X, Y and Z
dimensions, can explain any three-dimensional object such as a
box, sphere, planet, star, anything and also their meaningful
movements. All the objects you can observe are three
dimensional. The shape of objects and their movements can
only be imagined or observed relative to the dimensions X, Y, Z.
The fourth dimension is the Time "T" and this dimension is
beyond visual observation but the time dimension affectingly
effects everything's emerging existence; everywhere;
everywhen. For now, let the time dimension go and only
imagine the spatial dimensions as void space; like when you
close your eyes.When you close your eyes; everything is gone
and you are imagining infinite darkness everywhere; like dark
void space of Cosmos. This infinite darkness is the built-in
screen of the marvelous mind and every living human has this
built-in screen; where you can creatively create; continuously
calculate anything. The infinite darkness of infinite imagination
is the freedom of infinity. If you remove all the less massive;
more massive matters; every emitting electromagnetic emission,
energy from Cosmos; the remaining is infinitely vast void
surrounding space of darkness like the empty screen of initially
infinite imagination.

Now, create three spatial dimensions "X, Y, Z" inside infinite imagination of your marvelous mind; even in reality, the dimensions are actually imaginary. All the dimensions are imaginary straight lines starting from one common point "O" and they are perpendicular to each other or in other words ninety degrees relative to each other. One imaginary linear line X, horizontally extending from O to the right and left sides of O; One imaginary linear line Y, horizontally extending from O to forward and backward of O; One imaginary linear line Z, vertically extending from O to upward and downward of O.

The X and Y representing an imaginary horizontal plane and Z representing an imaginary vertical length. The X, Y, Z imaginary lines are starting from a common point but continuously increasing because the imagination is infinite. Now you have an entirely empty creative canvas; constantly continuous Cosmos in your mind.

# STAR SYSTEM'S STRUCTURE

## Superbly Shining Star

Light is Photon. Imagine a photon, as a small shining point of light for now. A star continuously emits photons, particles of light. From a very far distance, a shining star can be imagined as a point of light like you see in the night sky. Now, start to move closer to that star. As you are moving closer and closer to that star, it's becoming larger and larger in size. Now, you are just only few kilometers away from that star; so you are observing a large shining sphere made of highly hot flaming fire. The surface of a star is highly hot; so the extreme environment on the surface is completely cataclysmic.

In reality, Stars are spherical structures of extremely hot gases; mostly Hydrogen and Helium; it's remaining in this spheroid form because of its own Gravity. Stars use fusion reactions; hydrogens fuse together to form helium and release enormous electromagnetic emission, energy. These fusion processes continuously occur inside a star to produce light. For now, it will be better if you just imagine cosmic objects only, because we will describe them in the later sections soon. You can imagine a star of any color but can you imagine color? Now, think of yellow and remember it; that star in your imagination is now a Yellow Dwarf star; endlessly emitting yellow light; this star in your imagination is the Sun. The Sun is a yellow dwarf star; a greatly giant cosmic object relative to the Earth. The Sun's gravity is responsible for the gravitational orbit of the Earth and the other planets, around the Sun.

# Prominent Planet

A planet is a gaseous or sphere shaped solid object orbiting a star; like the Earth. The Earth is a special planet because it's perfectly habitable for lovely life; especially human life. Most planets of the universe are not habitable for life relative to the total number of planets in the universe. If I tell you to imagine a planet; what comes in your mind first? For me, it's the planet Earth our own home planet. Keep in mind that in these initial sections I will mostly describe shapes of astronomical objects with simple details; so that you can create those objects in your infinite imagination. After discussing prerequisite physics, I will describe more about those objects to make you understand better.

# Marvelous Moon; Natural Satellites

A moon is a spherical shaped solid structure, satellite orbiting a planet naturally. Moons are smaller relative to planets. If I mention moon; the moon of our home planet Earth is created in your infinite imagination. All the natural satellites, moons of the universe can be imagined relative to the moon of the Earth; so you can create any moon of the universe in your infinite imagination; relative to the moon of the Earth.

# Astronomical Asteroid

Astronomical Asteroids are randomly shaped; small sized solid star system's ordinary objects; some are spherical; some are not. Asteroids are smaller objects relative to planets and moons but larger astronomical asteroids are also actively advancing around stars and they are called planetoids.

# Cool Cosmic Comet

Comets are also randomly shaped; small sized solid star system objects but mostly icy. The difference between comet and asteroid is that a comet has an altering atmosphere around its body but it's mostly gravitationally unbound. Comets have high range of orbital periods and they also have highly eccentric elliptical orbits around stars. When comets are passing by close to stars, their icy elements become warm and release gas; that released gas creates an amazingly artistic atmosphere; appearing as an artistically astonishing; vastly visible tremendous tail.

# Remarkable Round Ring

Many giant planets in the cosmos have a planetary ring around them. From the name you can understand that the planetary rings are visually round rings or disk-shaped astronomical assortment around an astronomical object such as a planet; these planetary rings are composed of small sized solid materials such as dust, ice, rocks and also moonlets.

**Other Objects**
There are other objects in a star system such as meteoroids, minor planets, Planetesimal. Meteoroids are small sized solid solar system object; smaller than asteroids; mostly rocky or metallic. Minor planets directly orbit a star but they are nor planet neither comet. Planetesimals are small materials mainly wandering within debris disks around any shining star.

# Single Star System

Star systems are complexly combined cosmic collections of ordinary objects orbiting shining stars; within a star system, cosmic objects affected by that star's gravity are orbiting around it because of gravitational attraction. All the cosmic objects described above exist within significant star systems; they are gravitationally bounded by stars. Together, all those objects can create a significant star system but it's not necessary for a star system to contain all those objects. The most necessary condition of a star system, is that it must always contain one star but a star system can contain more than one shining star too; there are star systems in the cosmos, without any star system object within them such as planets, moons and asteroids. Now, creatively imagine the Sun in the common point of the imaginary dimensional system of your mind or imagine the sun in the center of your empty canvas inside infinite imagination.

Congratulation, you successfully created a simply stable single star system in the continuously creative cosmos of your independently infinite imagination; existing within your marvelously meaningful mind. Is this our own Solar System? No, it's not because our own solar system has the Sun with multiple planets, moons and asteroids; they all are orbiting around their shining star, Sun.

"Wise we, who undoubtedly understand our own observable Solar System brilliantly better than the other objects of surrounding spacetime; so we will completely create our own single star system, significant solar system inside imaginary dimensional diagram; designless dark canvas; currently creationless; completely clear canvas; virtually void; vacantly spanning stage of outstanding cosmic creationist's consciousness by brilliantly Implementing Infinite Imagination; the tremendously honorable humans are artistically creative cosmic consciousness; clever cosmic creationists."

# Shining Sun's Superb Solar System

Now, move seventy million kilometers away from the Sun and imagine the first planet of the solar system Mercury; it is orbiting the sun inside infinite imagination of your marvelous mind. Meters are the standard unit to measure spatial dimensions but this is your imagination, so you can take kilometers as measuring unit for now but soon we will use larger units.

Mercury is the smallest planet in the Sun's solar system. A planet or a star can be imagined as a spherical structure or a ball; If you move far away then that sphere will become significantly smaller and if you move closer to it then that sphere will become lively larger. Even imagination needs relativity; that's why I said that the relativity is a universal truth. When I say stars or planets; you imagine spherical shapes and relative to the spherical shape, those cosmic objects become reality in your infinite imagination. The Sun's star system contains eight planets and they are Mercury, Venus, Earth, Mars, Jupiter, Saturn, Uranus, Neptune and a notable minor planet, Pluto. Most of these planets have one or more moons.

# Muggy Mercury

The Sun's star system's spherically solid; significantly smallest planet is known as Mercury and it is the first planet; its diameter is only 4,879 kilometers. The orbital distance of Mercury from the sun ranges from 46 million to 70 million kilometers; so the mercury takes approximately 88 Earth days to complete one orbit around the Sun. The mercury has a solid inner core; a deeper hot liquid middle core, layer surrounding that inner solid core; then an iron sulfide solid outer core, layer; then a solid silicate crust and mantle overlying that solid outer core. The surface temperature of mercury is -173 degrees to 427 degrees Celsius. The Mercury's gravity is not enough to sustain significant atmosphere around it; it has an exosphere containing hydrogen, helium, oxygen, sodium, calcium and potassium. The planet Mercury has no natural satellite, moon. The Mercury's surface appears dark gray in color. Existing extraterrestrial evolution of life is impossible in the planet Mercury.

# Violent Venus

The Sun's star system's spherically solid; significantly smallest; second planet is known as the Venus; its diameter is only 12,104 kilometers. The average distance of Venus from the Sun is 108 million kilometers; so it takes 243 days to complete one orbit around the Sun. The size and mass of Venus are similar to the Earth and it has a rocky body like the planet Earth. The inner structure of Venus is also significantly similar to earth; it has a solid inner core; then a hot liquid middle core; then an overlying solid mantle and crust. The Venus has a dense atmosphere containing 96.5 percent carbon dioxide and some nitrogen and sulfur dioxide. The planet Venus has no natural satellite, moon. The surface of the Venus appears yellowish white in color. Existing extraterrestrial evolution of life is impossible in this planet.

# Excellently Extraordinary Earth; Humanity's Heavenly Home

The Blue Earth is the third planet of the Sun's star system; it is the only known perfectly habitable planet to humanity that endlessly ensured exceptionally essential evolution of lovely lives. Its diameter is only 12,742 kilometers. The Elegantly eminent Earth formed about 4.5 billion years ago. This perfectly habitable home planet is the only heavenly home; hosting honorable humanity.

This planet is orbiting the Sun from the average distance of 150 million kilometers; it takes 365.25 days to complete one orbit around its superbly shining stunning star, Sun. The Earth has only one obviously observable; overwhelmingly outstanding; orderly orbiting natural satellite named the moon. The Earth is my home planet and our only home; so, I dedicated a significantly special section for the "heavenly honorable home, endlessly essential eminent Earth."

# Marvelous Mars

Mars is the fourth planet and it is also the second smallest; spherically solid profound planet. Its diameter is only 6,779 kilometers. The distance of Mars from the Sun is 230 million kilometers and it takes 687 days to complete one orbit around the Sun. The planet Mars has an inner core containing iron, nickel and sulfur; this inner core is surrounded by mantle mostly containing silicate; the mantle is surrounded by crust containing oxygen, silicon, iron, magnesium, aluminum, calcium and potassium. The Color of Mars surface is red. The Mars has a less dense atmosphere mainly containing 96 percent carbon dioxide and argon, nitrogen, oxygen; it cannot contain liquid water because of its lower atmospheric pressure but it has two permanent polar ice caps containing frozen water. This planet has two small natural satellites like the Earth's marvelous moon; they are Phobos, Deimos. Existing extraterrestrial evolution of life is impossible in this planet.

# Astonishing Asteroid Belt

The Asteroid belt is the boundary of the inner solar system; it's a roundly surrounding region of the solar system; existing between the orbit of Mars and Jupiter. These astronomical Asteroids are located in the space between Mars and Jupiter. This disk-shaped belt contains small and large-sized Asteroids and a minor planet. The four largest Asteroids of the Asteroid belts are Ceres, Vesta, Pallas, and Hygea; they contribute half of the mass. Ceres is the only dwarf planet of the Asteroid belt; it is 950 kilometers in diameter; the diameters of other three named Vesta, Pallas, Hygiea are less than 600 kilometers.

**Hilda Asteroids**

This is the final boundary of Asteroid Belt; this boundary ends exactly before the orbital path of the Jupiter. Approximately four thousand large sized asteroids encirclingly existing within this region of Asteroid belt; they all are called Hilda Asteroid. The most unique feature of this area is that it has an unpointy triangular shape; if you observe the total area of Hilda Asteroids, it will appear to you as elliptically triangular in surrounding shape. A notably large asteroid of this region is 153 Hilda; this is also a minor planet with a diameter of 170 kilometers.

# Jumbo Jupiter

The fifth planet Jupiter is the largest planet in the Sun's star system. Its diameter is only 139,820 kilometers. This planet is a gas giant. The average distance is 778 million kilometers from the Sun; so it takes 11.86 Earth years to complete one orbit around the Sun. The Jupiter contains a dense inner core mostly made of mixtures of elements; then a middle layer of metallic liquid hydrogen with small amount of helium surrounding the inner core; then an outer layer surrounding the middle layer that mostly contains molecular hydrogen. A Gas Giant is an aesthetical astronomical apparatus that failed to become an especially shining star but never failing to hugely humor highly honorable humans. Jupiter has the largest atmosphere relative to all the planets in the solar system. The most extraordinary feature of this planet is its Great Red Spot; it's a continuous anticyclone storm larger than the entire Earth. This planet has seventy-nine natural satellites; some of them are Callisto, Europa, Ganymede, Io and they are also known as Galilean Moons. The Jupiter looks like a whitely orange, brown planet. Existing extraterrestrial evolution of life is impossible in this planet because it's a gaseous planet.

**Jupiter Trojans**
A large group of astronomical asteroids; encirclingly existing in Jupiter's orbit and also sharing Jupiter's orbital path.

# Stunning Saturn

The Saturn; Sun's star system's sixth spherical shaped superb structure; second largest gaseous planet; it's orbiting the Sun from an average distance of 1.4 billion kilometers; so this planet completes one orbit in 29.5 years. Its diameter is 116,460 kilometers. The Saturn has a small solid inner core like the Jupiter; surrounded by mostly hydrogen and helium. The Saturn atmosphere contains 96 percent molecular hydrogen and 3.3 percent helium. This planet has 62 natural satellites and some of them are Titan the largest, Rhea the second largest, Lapetus, Dione, Tethys, Mimas, Enceladus. The Sun's star system special Saturn; surrounded by its revolutionary rare roundly rotating ring; rightly revolving around it like a dreamingly designed; distractingly delightful; dauntlessly dancing disk. The color of Saturn appears yellowish brown. The Ring of Saturn extends 6,630 to 120,700 kilometers outward from its equator and it's only 20 meters in thickness; this ring contains its 62 moons and many small solid ice fragments and few other elements. Existing extraterrestrial evolution of life is impossible in this planet because it's a gaseous planet.

# Unique Uranus

The Sun's star system seventh spherical planet the unique Uranus is also a gas giant but significantly smaller relative to other outer solar system planets; its diameter is only 50,742 kilometers. This planet is orbiting the sun from an average distance of 3 billion kilometers; it takes 84 years to complete an orbit.

The unique future is that this planet's axis is tilted more than ninety degrees; it is the only planet that orbits the sun on its side. The Uranus has a solid inner core consisting of silicate, iron, nickel; surrounded by an icy middle mantle and a gaseous outer layer containing hydrogen and helium. This planet has 27 natural satellites and some of them are Ariel, Miranda, Oberon, Titania, Umbriel. The observed color of this planet is white blue. Existing extraterrestrial evolution of life is impossible in this planet because it's a gaseous planet.

# Noteworthy Neptune

The eighth planet of the solar system is Neptune; the farthest planet and significantly smaller relative to other outer solar system planets. The average distance of Neptune from the Sun is 4.5 billion kilometers; it takes 165 earth year to complete one orbit. The Neptune is similar to its twin sister Uranus; The diameter of Neptune is 49,244 kilometers. The observed color of Neptune is blue; its structure is similar to Uranus. This planet also has a planetary ring containing icy particles and also fourteen natural satellites; some of them are Triton the largest one, Nereid an irregular moon, Proteus, Naiad, Thalassa, Despina, Galatia and Hippocamp. Existing extraterrestrial evolution of life is impossible in this planet because it's a gaseous planet.

### Neptune Trojans
A large group of astronomical asteroids encirclingly existing in Neptune's orbit and also sharing Neptune's orbital path.

## Centaurs

These are sun's star system's solid small bodies encirclingly existing in the region of outer solar system planets; all of them usually have unstable orbits except only one Centaur named BZ. The Centaurs share the characteristics of both Asteroid and Comet.

# Kuiper Belt

The Kuiper Belt exists within Trans-Neptunian region; its area simply starts after the orbit of the Neptune and then extends to 7.48 billion kilometers from the sun. This belt is actually a continuous circumstellar disk beyond the outer solar system region. It's similar to Asteroid belt but greatly larger in size and mass; containing small solar system bodies; resulting remnant of the solar system formation. Most Kuiper belt ordinary objects are composed of frozen water, methane, ammonia and many asteroids are also composed of rocks and metals. The three major minor planets Pluto, Haumea and Makemake exist in this region; they are also called dwarf planets.

**Scattered Disc:** Icy solid small solar system bodies existing beyond Neptune's orbit. Border of the Trans-Neptunian region starts from here. The Scattered Disc overlaps the Kuiper belt; it mostly contains comets and icy small fragments.

**Pluto:** This is the largest known dwarf planet in the Kuiper belt region; it's made of ice and rock. The orbital distance of Pluto ranges from 4.4 to 7.4 billion kilometers; so it takes 247.7 years to complete one orbit around the Sun. The Pluto has one natural satellite named Charon. Existing extraterrestrial evolution of life is impossible in this dwarf planet.

**Haumea and Makemake:** These are also minor planets existing in Kuiper Belt but they are significantly small sized and less massive relative to Pluto.

# Heliosphere

The heliosphere is the final boundary of the solar system; it acts like a bauble shield that protects the solar system's inside area, from the interstellar medium. This Stellar wind bauble radiates sun's solar wind at 400 kilometers per second. The Solar wind, streams of charged particles intentionally interacts with winds of interstellar medium constantly to protect solar system objects.

**Distant Detached Ordinary Objects:** Trans-Neptunian minor planets orbiting the Sun from the outer reaches of the solar system; they are also called sednoid. These minor planets orbit the sun from a distance greater than any other ordinary objects; so they are classified as detached star system objects.

**Gravitational Field Boundary:** The Sun's gravitational field's dominant effects end at the distance of almost two light years from the center of the Sun; this is the Sun's star system's gravitational boundary. The light year is a larger unit than kilometers; one light year is approximately 9.461 trillion kilometers.

I initially described all ordinary objects of the Sun's star system; so it's time to create this significant star system inside infinite imagination of your mind. Once you successfully shape Sun's star system inside infinite imagination of yours; you will also be able to imagine other star systems relative to this significant Solar System.

# Implementing Infinite Imagination

Close your eyes to imagine the empty imaginary dimension system like dark vast void space; then create the Sun in the center of your imaginary dimension system; this is a simple star system with only one star; then create the first planet Mercury in the first orbital position of the Sun; then the second planet Venus in the second orbital position; then the third planet Earth in the third orbital position with only one moon; then the fourth planet Mars in the fourth orbital position with two moons; then the Asteroid belt containing millions of Asteroids between Mars and Jupiter; then the fifth planet Jupiter in the fifth orbital position with seventy-nine moons; also keep imagining Centaurs from the position of Jupiter till Neptune; then the sixth planet Saturn in the sixth orbital position with sixty-two moons orbiting the Saturn within its large disk shaped Planetary Ring; then the seventh planet the Uranus in the seventh orbital position with twenty-seven moons orbiting the planet within its disk shaped small Planetary Ring; then create the eighth planet Neptune in the eighth orbital position with fourteen moons orbiting the planet within its disk shaped small Planetary Ring. The major planets are created in your cosmic imagination; now you will go beyond that.

After the position of Neptune create the scattered disc, described before and also the Kuiper Belt; then the dwarf planets Pluto with one moon, Haumea and Makemake. After all of those, create the bauble shield, Heliosphere the final boundary of a star system. Congratulation, using the power of infinite imagination you created a greatly gigantic solar system within your mind.

"Once you discover the power of infinite imagination you achieve infinite independence; fathom fearless freedom of your marvelously magnificent mind that can invent infinite inventions; conquer constant complexities; survive stressful struggles; smartly solve puzzling problems; we will wander beyond brightly shining Stars soon."

# GENERATING GALAXIES

"Greatly glowing galaxies are compactly combined complex collections of separated star systems; bondingly bounded by galactically governing gravity; generally guiding shining star's elliptical encircling; several star systems affected by their gravities generate greatly gargantuan galaxies."

Milky Way, the home galaxy of humanity containing the Sun's star system and also the Earth. Star systems are main components of a Galaxy but it also contains many more objects than just star systems and we will discuss them in this section. Different galactic objects together form galaxies; after acquiring important information of galactic objects; you will be able to generate galaxies in your infinite imagination. In the end of this section, we will describe the structure and motion of a galaxy in a unique alphabetical alliteration of only "C".

## Shining Star's Stratification

## By Surface Temperature

Star types are classified by their electromagnetic radiation's spectral line; stars will be discussed more after the physics section. For now, I will describe star types based on their surface temperature. The main types are classified using letters O, B, A, F, G, K, M, L, T, Y based on that star's surface temperature.

**"O" Type:** Surface temperature of O type star is equal to 33,000 Kelvins or more; This type of star is the most massive and luminous in the Unlimited Universe.

**"B" Type:** Surface temperature of B type star is 10,000 to 33,000 Kelvins; This type of star is less luminous and also less massive than "O" but more than "A".

**"A" Type:** Surface temperature of A type star is 7,500 to 10,000 Kelvins; This type of star is less luminous and also massive than "B" but more than "F".

**"F" Type:** Surface temperature of F type star is 6,000 to 7,500 Kelvins; This type of star is less luminous and also less massive than "A" but more than "G".

**"G" Type:** Surface temperature of G type star is 5,200 to 6,000 Kelvins; This type of star is less luminous and also less massive than "F" but more than "G".

**"K" Type:** Surface temperature of K type star is 3,700 to 5,200 Kelvins; This type of star is less luminous and also less massive than "G" but more than "M".

**"M" Type:** Surface temperature of M type star is 2,000 to 3,700 Kelvins. This type of star is less luminous and also less massive than "K" but more than "L".

**"L" Type:** Surface temperature of L type star is 1,300 to 2,000 Kelvins; This type of star is less luminous and also less massive than "M" but more than "T".

**"T" Type:** Surface temperature of T type star is 700 to 1,300 Kelvins; This type of star is less luminous and also less massive than "L" but more than "Y".

**"Y" Type:** Surface temperature of Y type star is equal to 700 Kelvins or less; This type of star is the least luminous and also least massive in the Unlimited Universe.

All these letter types also have sub-divisions from 0 to 9 based on their properties or order of decreasing temperature; such as the type A's (10,000 – 7,500) = 2,500 is divided into 0 to 9 or ten divisions; 0 indicating highest temperature and 9 indicating lowest temperature of that particular main type; "0 – 9" is "High to Low" in temperature.

# By Luminosity and Size

The third classification is based on the star's spatial size and luminosity.

**Type "0":** Extremely Luminous Supergiant Stars are in type 0.
**Type "Ia":** Normally Luminous Supergiant Stars are in type Ia; relative to type 0.
**Type "Ib":** Less Luminous Supergiant Stars are in type Ib; relative to type Ia.
**Type "II":** Bright Giant Stars are in type II; less massive relative to type Ib.
**Type "III":** Normal Giant Stars are in type III; relative to type II.
**Type "IV":** Subgiant Stars are in type IV; relative to type III.

**Dwarf Stars:**

**Type "V":** Main Sequence Stars or Dwarf Stars are in type V.
**Type "VI":** Sub Dwarf stars are in type VI; relative to type V.
**Type "VII":** White Dwarf stars are in type VII or D.

The Main Sequence Stars, small sized stable stars will be discussed in a different section. The Sun is a G2V type star; "G2" indicating that its surface temperature is 5,778 Kelvins and "V" indicating that it's a Main Sequence Star. All those types of stars exist within galaxies.

# Actual Apparent Color

**Blue:** Most of the "O" type stars are Deep Blue in apparent color.

**Deep Blue White:** "B" type stars are Deep Blue White in apparent color.

**Blue White:** "A" type stars are Blue white in apparent color.

**White:** "F" type stars are White in apparent color or yellow-white.

**Yellow:** "G" type stars are Yellow in apparent color.

**Orange:** "K" type stars are Orange in apparent color.

**Red:** "M" type stars are Red in apparent color.

**Red Brown:** "L" type appears Red Brown in the Stellar spectrum but it will appear as Scarlet to the human eyes.

**Brown:** "T" type appears Brown in the Stellar spectrum but it will appear as Magenta to the human eyes.

**Dark Brown:** "Y" type appears Dark Brown in the Stellar spectrum but it will appear as Purple to the human eyes.

Star colors changes from Blue to White to Yellow to Orange to Red as the temperature decreases from maximum to minimum.

# Star Types by Size & Color

## Main Sequence Star; Dwarf

### Red Dwarf
"M" type Red color main sequence stars; small sized relative to the Sun; star mass is 0.5 to 0.8 solar masses. One solar mass is equal to the mass of one Sun.

### Yellow Dwarf
"G" type Yellow color main sequence stars; similar sized relative to the Sun; similar star mass relative the Sun.

### Orange Dwarf
"K" type Orange color main sequence stars; small sized relative to the Sun; star mass is 0.5 to 0.8 solar masses.

## Special Dwarf

### White Dwarf
This star is actually a white core of a small sized star that cannot become a neutron star with its insufficient mass. Its mass is similar relative to the Sun; size similar to the Earth.

### Black Dwarf
A black dwarf is a very cold white dwarf that cannot emit any heat or light radiation.

**Brown Dwarf**
These brown color small astronomical objects are neither stars nor planets. Its mass is similar to Jupiter.

**Blue Dwarf**
These are the theoretical remnants of Red Dwarf Stars; very small sized.

**Ultra Cool Dwarf**
"M" type cold red dwarf star; relative to other ordinary Red Dwarfs.

# Sub-Dwarf

Subdwarfs are less massive and small sized relative to normal dwarf stars. There are hot, cool and brown sub-dwarfs.

# Giant Star

Giant stars are larger in size and greater in mass relative to Main Sequence Dwarf Stars; these stars are few hundred times larger in size than the Sun; almost ten to thousands times luminous relative to the Sun; the mass range is between seven to hundred times of the Sun.

**Blue Giant**
Blue color giant stars with huge mass and high luminosity relative to the Sun.

**White giant**
These stars are actually Blue Giants; appearing white in color because of their low temperature relative to common Blue Giants.

**Red Giant**
Red color giant stars with huge mass and high luminosity relative to the Sun.

**Yellow Giant**
Yellow color giant stars with huge mass and high luminosity relative to the Sun.

**Bright Giant**
These are special type of Giant Stars with more luminosity relative to normal Giant Stars but less than Supergiant Star.

# Sub Giant Star

These are special type of stars larger in size with more mass and luminosity than main sequence stars but less than Giant Stars.

# Supergiant Star

These Stars are the largest and most massive stars in the common classification of stars. This type of stars has highest surface temperature and also has highest luminosity. These stars are thousand times larger in size and also hundred times massive relative to the Sun; the luminosity range is maximally million times to ten thousand times than the total luminosity of Sun.

**Blue Supergiant Star**
Blue color extremely luminous Supergiant Stars.

**White Supergiant Star**
Especially, Blue Supergiant Stars but always appear as white-blue to observers.

**Red Supergiant Star**
Red color largely luminous Supergiant Stars.

**Yellow Supergiant Star**
Yellow color largely luminous Supergiant Stars.

# Hypergiant Star

These are remarkably rare; marvelously massive; superiorly shining stars of the Unlimited Universe. These stars have hugely hot high temperature; they are even more luminous relative to supergiant stars. These stars are almost one to three thousand times larger in size relative to the Sun.

**Blue Hypergiant Star**
Blue color extremely luminous Hypergiant Stars.

**Red Hypergiant Star**
Red color extremely luminous Hypergiant Stars.

**Yellow Hypergiant Star**
Yellow color extremely luminous Hypergiant Stars.

All those stars exist within galaxies.

These topics are mostly related to Stellar Evolution; it will be discussed more after the physics section.

# Star System's Stratification

## Single Star System

A star system that contains one star of any type is a single star system. This star system can contain only one star or a star with other objects such as planets, moons, minor planet and asteroids. The Sun's star system is a Single Star System.

## Binary Star System

A star system that contains two stars of any type is a Binary Star System. This star system can contain only two stars or two stars with other objects such as planets, moons, minor planets and asteroids. In a binary star system, two stars orbit themselves because of their total gravity.

## Trinary Star System

A star system that contains three stars of any type is a Trinary Star System. This star system can contain only three stars or three stars with other objects such as planets, moons, minor planets and asteroids. In a trinary star system, three stars orbit themselves because of their total gravity.

# Quaternary Star System

A star system that contains four stars of any type is a Quaternary Star System. This star system can contain only four stars or four stars with other object such as planets, moons, minor planets and asteroids. In a quaternary star system, four stars orbit themselves because of their total gravity.

Star systems containing more than four stars can also exist; they are Quintenary Star System, a five-star system; Sextenary Star System, a six-star system; Septenary Star System, a seven-star system; or maybe more.

Those star systems exist within galaxies.

# Proper Planetary Presentation

# Common Classification

**Gaseous Planet:** A planet with a solid inner core surrounded by massive gaseous structures, such as Jupiter, Saturn, Uranus, Neptune.

**Ordinary Solid Planet:** A planet with solid spherical structure but not gaseous at all and also not habitable for life, such as Mercury, Venus, Mars.

**Habitable Planet:** Habitable planets existing in the habitable zone of its star system; their properties are significantly similar to the Earth.

# Classification by Mass

**Giant Planet:** A planet which is massive; mainly made of Hydrogen and helium Gas or water, methane, ammonia, ice or rarely rocks.

**Mesoplanet:** Planets, smaller than Mercury but larger than Ceres; their size is one thousand to five thousand kilometers in diameter.

**Mini Neptune:** Gaseous planets smaller than Neptune and Uranus; mass is up to ten Earth's masses and also called gas dwarf.

**Super Earth:** A planet outside sun's star system; more massive than Earth but less than Neptune and Uranus.

**Super Jupiter:** An astronomical object more massive than the Jupiter.

**Sub Earth:** Planets less massive than the Earth and Venus.

# Planetary Mass Objects

Astronomical objects massive enough to sustain spherical shapes by their own gravities but not massive enough to start fusion to become stars are called Planetary Mass Objects. All planets are planetary mass objects; some special types of cosmic objects also belong to this classification.

**Dwarf Planet:** A planetary mass object that is not natural satellite or planet; this object's gravity is not enough to clear other objects from its own orbital path around its star. The Pluto is a dwarf planet.

**Rogue Planet:** Any astronomical object ejected from its own star system into the interstellar medium; roaming in the deep space.

**Captured Planet:** Roaming rogue planets captured by any other star rather than its parent star.

**Satellite Planet:** Natural satellites; similar size or larger than mercury.

**Former star:** Sometimes in a close binary star system, the less massive star can lose its mass to its more massive companion and become a former star object.

# Compositional Classification

**Oceanic Planet:** A planet with most mass made of water.

**Desert Planet:** A planet with most of its surface covered by desert.

**Icy Planet:** A planet with most of its surface covered by ice.

**Lava Planet:** A planet with most of its surface covered by lava.

**Gaseous Planet:** A planet with massive gaseous outer layers.

**Iron Planet:** A planet mostly made of Iron.

**Terrestrial Planet:** A rocky planet mostly made of carbonaceous rocks, silicate rocks or metals.
**Chthonian Planet:** A extrasolar planet orbiting very close to its star that lost its gaseous outer layer because of the gravity of that star.

**Coreless Planet:** This is a theoretical planet without an inner core like a hollow planet.

**Habitable Planet:** This is actually a potentially habitable planet orbiting a star many light years away from the Sun. Intelligent life forms capable of infinite imagination and continuous creations can exist on that planet full of cities like the Earth; maybe those planets are eminently existing within Colossally Continues Cosmos.

# Orbital Classification:

**Exoplanet:** Planets orbiting other stars rather than the Sun.

**Extragalactic Planet:** Planets existing outside the Milky Way, human's home galaxy.

**Eccentric:** A giant planet with an eccentric orbit around its star. Orbital eccentricity is the measurement of roundness of that orbit. Zero eccentricity means more circular orbit; higher number eccentricity means more elliptical orbits.

**Binary Planet:** Two planets orbiting each other because of gravity.

**Circum-Binary Planet:** Planets orbiting binary stars or two stars.

**Circum-Trinary Planet:** Planets orbiting three stars. Planets orbiting multiple stars can be classified like this, such as Circum-Quaternary and so on.

**Hot Planet**: Planets orbiting very close to its star.

**Major Planet:** An astronomical object dominating its own orbital path and the majority mass of its position.

**Minor Planet:** An astronomical object failing to dominate its own orbital path and the majority mass of its position.

**Habitable Zone Planets:** Planets orbiting in the Habitable Zone of a star system.

# Planetar

These astronomical objects are not stars or planets but can be considered as very special small sized shineless starlike spherical structures; existing within magnificently mysterious macrocosm. Brown Dwarfs and Sab Dwarfs are actually Planetars. These objects may or may not orbit any star.

**Brown Dwarf**
Astronomical objects having mass between minimum massive small stars and maximum massive gas giants; these objects are different from real stars because they fuse Deuterium and Lithium rather than Hydrogen and Helium because of their insufficient low mass, relative to real shining stars.

**Sub-Brown Dwarf**
This astronomical object forms from gravitational gas cloud collapse like Brown Dwarfs and Stars but has less mass than Brown Dwarfs. Like brown dwarfs, sub-brown dwarfs cannot fuse Hydrogen and Helium; so they are also called planetary mass brown dwarf or freely floating planets of Unbound Universe; Expanding Entirety.

# OTHER OUTSTANDING OBJECTS

## Neutron Star

A neutron star is a high-density star of high mass relative to its small size; relative to other ordinary stars. This star is the remaining core of a giant or supergiant star after it goes supernova. Neutron stars are significantly small sized white dwarfs relative to other ordinary white dwarfs; because of their extreme gravity neutron stars can spin up to seven hundred times per seconds. Supernovae occur when giant stars lose the continuous capability of sustaining their own mass; so they collapse because of their own gravity; in other words, this type of explosion occurs when giant stars die dramatically. These extreme explosions are so colossally cataclysmic and amazingly luminous that their enormously extreme energy emissions enlighten up dark regions of the Unbound Universe.

"After the supernova of giant stars, the gravitational collapse compresses central cores of those shining stars beyond the density of ordinary white dwarfs to flawlessly form neutron stars; magnificently monumental marvels of marvelously mysterious macrocosm."

An average neutron star has a radius of ten to twelve kilometers but they are approximately as massive as one or two Suns. From the star structure model, the structure of a neutron star can be guessed but it's still a mystery; a neutron star may contain an inner core consisting of quark-gluon plasma; surrounded by an outer core consisting of neutron-proton Fermi liquid with few percentages of electron fermi gas; surrounded by an inner crust consisting of electrons, neutrons, nuclei; surrounded by an outer crust consisting of ions, electrons; then a surrounding surface possibly made of iron nuclei and free electrons. The gravity of a neutron star slows down time and can curve the photon's path to an observational level. Don't worry if you don't understand these texts now, because you will understand them after the physics section.

There are two types of Neutron Stars and they are Magnetar and Pulsar.

# Magnetar

Magnetars are actually neutron stars; these stars are magnificently magnetic with extremely powerful magnetic fields surrounding them. Magnetars decay magnetic fields to emit extremely enormous emissions of electromagnetic radiation, mostly x- rays and gamma rays. As it is a neutron star it also forms from a supernova explosion of a star; if the spin, temperature and normal magnetic field of that neutron star remain in the right range, it can create continuous dynamo mechanism to become a Magnetar.

# Pulsar

Pulsars are magnificently magnetized; rapidly rotating and always active neutron stars emitting extremely enormous emissions of electromagnetic radiation and also shooting jets of highly hot particles from their two magnetic poles to vast distances of Unlimited Universe. Pulsar's perfectly pulsing property classified it as a special neutron star called Pulsar; they are observed to pulse different types of electromagnetic radiation such as radio wave and light. Pulsars are one of the sources of Ultra High Energy Cosmic Rays (UHECR).

"Pulsars are the lovely leading lighthouses of constantly continuous cosmos."

# Nebula

Nebulas are the source of countless continuous creations of colossally continuous cosmos. A Nebula is an astronomical gigantic gas cloud of hydrogen, helium and other ionized gas and also dusts. Most of these gigantic gas clouds are almost hundreds of light years in size. Nebulas form from the enormous amount of existing gas of interstellar medium and also from the dust and gas ejected from extreme explosions of dying stars; the mutual gravitational attraction of gas and dust particles intensely increase their total density to forms new nebulas naturally.

"The denser divisions within nebulas relative to their less dense divisions frequently forms notably new shining stars; magnificently mysterious gravity governingly generates newly notable shining stars within wonderfully neat nebulas."

# Classification of Nebula

## Diffuse Nebula

These nebulas are randomly shaped nebulas with no defined boundaries; Most nebulas are Diffuse Nebula; they can contain considerably massive mass and also extend enormously. There are three typical types of Diffuse Nebulas.

### Emission Nebula

These diffuse nebulas mostly contain highly hot ionized gas; these nebulas emit electromagnetic emissions of various wavelengths. The significant sources of their initial ionization are high energy photons of enormously extreme electromagnetic emission emitted from nearby highly hot stars.

### Reflection Nebula

These diffuse nebulas are colossal clouds of interstellar dusts; they reflect emitted light of nearby stars but these stars do not emit enough extreme electromagnetic emissions or high energy photons to ionize them as emission nebulas.

### Dark Nebula

The dark nebulas are colossal clouds containing cold gas and significantly small sized dust particles; they cannot emit electromagnetic emission of visible light. The dark nebula hides its source star within its dominant darkness.

# Supernova remnant

These are significantly special types of Nebulas. When a star goes supernova its expanding layer of gas can form these significantly special diffuse nebulas called Supernova Remnant. There are mainly two types.

**Planetary Nebula** These are the remnant of final phase of relatively low mass star's stellar evolution process. These nebulas are relatively small sized with less mass than other nebulas.

**Protoplanetary Nebula** This nebula exists before the Planetary Nebula; its source is the central evolving star surrounded by this Nebula.

# Black Hole

A black hole is a marvelous cosmic object generating extreme gravitational effect; because of this extreme gravity, no particle or even light can escape it. A simple black hole has mass and gravitational effects. Its simple structure is that it contains a Spacetime Singularity surrounded by its inescapable Event Horizon. After supernova explosions of giant stars; the only resulted remaining are massive star cores but if those star cores have more mass than 2.2 solar masses; they compressingly collapse more because of their extreme gravity and then become Black Holes. It is very hard to observe a black hole in the Unlimited Universe because it does not emit or reflect light but Black Holes can be detected by observing their gorgeous gravitational effects; affecting surrounding spacetime, electromagnetic energies and all masses, matters.

# Schwarzschild Radius

The Schwarzschild radius is not the radius of a black hole but the distance of event horizon from its central singularity; the event horizon is a round region that exists around the singularity of a black hole; from there even light cannot escape with its great velocity because of the gravitational attraction of a black hole. In other words, the escape velocity is equal to the speed of light at the event horizon of a black hole; so photons of light fail to escape from that region. Centrally compressing any mass more beyond, internally inward its Schwarzschild Radius will cause a Black Hole of that mass; there is no radius of a black hole because the black hole is just a spacetime singularity. The Singularity is a 1-dimensional point of infinite gravity; it will be discussed more in the physics section.

"Black Holes are the most mysteriously magnificent; profoundly particular points; effectively existing within wonderfully worthwhile colossally continuous cosmos; constantly creating extremely enormous effects of governing gravitation; every energy's electromagnetic emission; every particle of matters fails to flee and also forcefully falls inside its Event Horizon."

# Black Hole Classification

# By Rotation and Electric Charge

There are four types:

**Non Rotating Uncharged Black Hole**
These are the most common Black Holes in the Unlimited Universe. This type of black hole does not rotate on its axis and also has no electric charge.

**Non Rotating Charged Black Hole**
These black holes are not common in the Unlimited Universe. This type of black hole does not rotate on its axis but always has electric charge.

**Rotating Uncharged Black Hole**
These black holes are rare in the Unlimited Universe. This type of black hole always rotates on its axis but has no electric charge.

**Rotating Charged Black Hole**
These black holes are also rare in the Unlimited Universe. This type of black hole always rotates on its axis and also always has electric charge.

# By Mass

## Stellar Mass Black Holes

This type of Black Hole forms from the gravitational collapse of a star after its supernova explosion; it has a minimum mass of three solar masses and a maximum mass of less than hundred solar masses. One solar mass is equal to the mass of the Sun.

**Micro Black Hole:** This is a tiny size hypothetical Black Hole; it's also called Mini Black Hole or Quantum Mechanical Black Hole.

## Interstellar Black Hole

These Black Holes are more massive than stellar mass black holes; this type of black hole has a minimum mass of one hundred solar masses and a maximum mass of one hundred thousand solar masses. The formation of these black holes is still a mystery but there are few predictions; they form from the merging of stellar black holes and by consuming common astronomical objects; it is also possible that they formed during the Big Bang.

# Supermassive Black Holes

These Black Holes are the most marvelous; most magnificently mightiest; most massive in the Unlimited Universe; this type of black hole is the only known single object in the Unlimited Universe that can capably contain this tremendously extreme; enormously massive mass. This type of Black Hole has a minimum mass of hundreds of thousands solar masses and a maximum mass of billions solar masses. Every massive gigantic Galaxy contains a central Super Massive Black Hole.

The formation of a supermassive black hole is still a mystery but there are predictions. It is accepted that when a black hole is in the central position of a Gigantic Galaxy, it stars to greatly grow by consuming surrounding matters; merging with other ordinary black holes; the gravity is strongest near the center; so most mass moves toward the center of a Galaxy.

One theory predicts that the black holes of tens or hundreds of solar masses resulting from hypernova of massive stars are the initial origins of Super Massive Black Holes; another theory predicts that before the formation of ordinary stars, massive gas clouds collapsed into significantly special stars called Quasi Stars because of gravity; these unstable stars can completely collapse directly into massive Black Holes. Another prediction is that closely combined clusters of stars formed massive Black Holes after their gravitational core collapse. Finally, it is possible that the Big Bang created all the Black Holes at the center of Massive Galaxies; then those Black Holes became Super Massive Black Holes by consuming surrounding matters and merging with other black holes.

# Quasar

A quasar is an extremely luminous; largely massive astronomical object existing within an active galactic nucleus. A galactic nucleus contains a supermassive black hole; so a quasar is also an active special supermassive black hole. When an accretion disk exists around a black hole it formally called a quasar; celestial collections of matters are unlikely to fall directly into a black hole but they fall while circling around it because of angular momentum; this particular process causes an accretion around an active black hole. An accretion disk is a spinning disk of extremely hot particles of gas; these gas particles continuously fall into the black hole and emit extremely enormous electromagnetic energy. Because of the emission of high energy, the quasars are the most luminous objects in the Unlimited Universe. Galaxy is not a single object but a celestial collection of astronomical objects; so a quasar is one of the most massive; magnificently luminous; large sized single outstanding object in Colossally Continuous Cosmos.

Almost all astronomical objects of galaxies are described simply, so that you can completely create them in your infinite imagination. I was a man of my words last time and also this time; so here's an alliteration of only "C" describing galaxies.

"Consistently continuous Cosmos containing colossally combined clouds; continuously creating countless colorfully candescent completely convex celestial cores; carrying countless completely convex colorless celestial creations; continuously circling colorfully candescent completely convex cores; completely circumnavigating complexly combined colorful constellations; containing central caliginous cataclysmic canyons; can constantly consume closely circling compact circular cloud containing countless cosmic components; colorfully candescent cores; celestial creations coming closer; crossing completely constraining circumference."

Simply meaning Cosmos containing massive gas clouds and those gas clouds continuously creating colorfully shining stars and those stars carrying spherical planets and planets are incapable of creating colorful lights and those planets continuously orbiting their colorfully shining stars and those stars with their planets completely orbiting within galaxies around galactic centers that containing collections of massive stars and a central black hole and a black hole can consume countless objects within its closely circling accretion disk  and stars and any astronomical objects when they orbit it very closely or come close to it and cross its dominant gravitational boundary, the event horizon.

# Infinite Imagination's Infinite Implementations

In here, we will discuss the infinite capabilities of imagination. To go beyond your current creational capability and informative intelligence you must understand the infinite power of human imagination. It is the power of imagination that can give you supreme success. The infinite imagination gave humanity the capabilities to achieve significant scientific breakthroughs and also established the current civilization of humanity. This is scientific literature so we will mostly discuss scientific information in here but remember that infinite imagination is beyond science or any actuality.

"Implementing infinite imagination is the ultimate understanding and also a universal use of cosmic consciousness; can creatively give heavenly honorable humanity the compulsory continuous capabilities to endlessly explore huge heavens of the Ultimately Unlimited Universe."

## Infinite Creation

The most powerful capability of infinite imagination is Infinite Creations of infinite sizes, shapes, quantities and properties. Understanding Unlimited Universe can immensely increase imaginative capabilities of all kind because the more intellectual information you gain; the more it increases infinitely imaginative creational capabilities. Imagine two blue marbles of same sized spherical shape; now increase the second one in size, so that the first marble appears a small sized point relative to the second one; congratulation you applied relativity within imagination.

Now, change the properties of the first marble to the blue Earth and also change the properties of the second marble to a blue glowing star; then put the blue earth on orbit around the blue star. Congratulation, you created an entirely new star system in your infinite imagination. Now, change the properties of the blue star to the Sun and also create seven blue marbles; change the properties of those marbles to planets; finally, you have a solar system of eight planets similar to Sun's solar system. You can change a marble to a dark marble and then put a glowing ring of light around it to create a Black Hole. Changing marbles is not the only way but you can use different objects of different shapes and sizes relative to the objects you want to create; then change them to their final forms. Its Infinite Imagination's Infinite Independence.

# Infinite Action

The most important capability of infinite imagination is the ability to do multiple actions at the same time. Imagine two marbles and change them to two neutron stars; now collide them and while they are colliding put a new star in an orbit around them. You can put multiple astronomical objects in orbital motion around a black hole or star or any massive body; you can also collide few objects within this system if you want to. At first, create one black hole and multiple objects around it in a safe distance; then apply actions to those objects to give them orbital motions around that black hole; Now, create another two black holes near the first one then apply actions to collide and merge them; three black holes are merging while many astronomical objects remaining in orbital motion around them; then create the gravitational waves, spacetime waves of merging massive black holes shaking system's every object.

# Infinite Speed

Within Infinite Imagination a cosmic consciousness, you can go faster than the speed of light and also achieve awesome speed, infinite speed. Create the sun's star system in your imagination move near the Sun and you are there; then go near the Neptune and you are there as soon as I mentioned "Neptune". Apply Infinite Creation to create a trinary star system at a distance of 4.36 light years away from the sun; now apply infinite actions to keep two relatively larger stars of them in orbit and also the third relatively small star in orbit around those two stars; this is the Alpha Centauri. Infinite imagination's infinite speed gave you the capability to travel 4.36 light years within less than a second when just I mentioned "Alpha Centauri". Using Infinite speed of infinite imagination, you can even escape a black hole or go beyond its event horizon to explore the supreme singularity.

# Infinite Independence

Infinite imagination is Infinite independence; you can do everything in here; you can break the laws of the cosmos; combine combinable cosmic creations or unlimited uncombinable units or shapes or laws or anything. Your infinite imagination is your infinite independence; your rules; your laws; your freedom from fairly everything.

# Infinite Position

One of the extraordinary power of infinite imagination is infinite positioning of cosmic consciousness, human. This is not just a power of imagination but the reality of subatomic particles; these particles can exist in multiple positions in same time; human's brightly brilliant brain containing cosmic consciousness is also made of subatomic particles; wonderingly wandering within waves. Create multiple star systems with many planets at the same time in different imaginary screens, sections of your infinite imagination and claim your positions in all those systems and also explore their particular planets at the same time; then create another imaginary screen section and explore a black hole or anything at the same time. So remember that you have a brilliant biological brain, an advance and infinitely intelligent; excellently eminent; electromagnetically extreme; qualitatively questioning quantum computer; continuously calculating; completely creating constant colossal cosmos.

# Infinite Perception

Infinite Imagination can give you Infinite Perceptions of infinite objects and actions. Mainly it's the capability to zoom in and out within your imagination; you can observe the solar system from inside or from far away; you can observe the sun from the surface or go beyond its surface to observe its fusion core.

You can also zoom in inside an atom to observe its structure. You can zoom in to the nucleus of a galaxy to observe its central supermassive black holes and its extremely luminous surrounding or zoom out to observe the entire galaxy from above or from sides. The truth is that you can observe objects of the cosmos from different positions or angles or ways to understand it differently.

# Infinite Possibility

Actions of objects can have infinite possibilities in infinite imagination. You can create infinite possibilities; predict their results. You can create the sun's star system; then erase the sun from there or change its gravity to antigravity, repulsive energy to observe its resulting situations; the ejections of all planet outside the solar system into the interstellar medium. You can imagine the final possibilities of the universe or existence of many more magnificent macrocosm; marvelously mysterious Multiverse.

# Greatly Glorious Galactic Imagination's Implementations

The Milky Way, humans heavenly home galaxy, a greatly gigantic galaxy relative to any star system; this galaxy alone approximately contains three hundred billion shining stars; we can't even count this total number from its star to end. A galaxy cannot be created as a star system within imagination but the infinite imagination can create billions of star systems instantly; a galaxy is mainly a compactly combined collection of separated star systems; so you only need the initial information of galactic size, shape, structure and motion to generate gigantic galaxies inside infinite imagination. The milky galaxy is approximately 190,000 light years in diameter, one end to another; even light will take 190,000 years to travel one end to another of this gigantic galaxy but you can completely generate gigantic galaxies like this, inside infinite imagination instantly if you want. The shape of Milky Way galaxy is elliptical; this is a barred spiral galaxy; so its simple structure is that it contains central bar-like structure with four curve spiral arms curvingly coming out from its large luminous central core; containing one supermassive black hole and few interstellar black holes with several stars orbiting around them. The mass of the supermassive black hole in the center of Milky Way is approximately 4.3 million solar masses; the diameter of its event horizon is 25.4 million kilometers.

Now, create this supermassive black hole in your infinite imagination; then create few interstellar black holes orbiting around it; then also create serval supergiant shining stars and many giant stars and asteroids orbiting it; in there many macrocosmic objects are also colliding continuously; causing cataclysmic catastrophes. Finally, this is the extremely luminous center of Milky way inside infinite imagination; then create four shining spiral arms curvingly extending outward from the galactic central bar; these supersized spiral arms contain countless shining stars; then give it all an elliptical shape. Finally, this is your own Milky Way galaxy relative to the Milky Way of the Universe; the Sun is almost 27,000 light-years from its center. You can explore this Galaxy using infinite imagination's infinite implementations; so visit the sun's star system; visit several star systems; zoom in or zoom out; visit its central core; it's infinitely independent imagination of only ours.

"Infinite Imagination, profoundly pioneering power of Cosmic Consciousness of Honorable Humans; this profound power can continuously create scientific success; legendary literatures and any amazingly artistic astoundingly attractive awesome Art and any important ideas; inventing inventions; establishing essential economical evolutions; encouraging economists, entrepreneurs, engineers, employers, designers, doctors, agronomists, artists, students, scholars, scientists, society, lawyers, leaders, humanity, humans, Creative Cosmic Civilians; Complete Cosmic Civilization."

# PERFECTLY PROVEN PHYSICS

In here, perfectly proven physics is explained without mathematics.

## Classical Mechanics

This is one of the oldest section of physics; it's mostly describing dynamics, collision and forthcoming future of ordinary objects when forces act on them. Simple gravitational motion can be described using classical mechanics; any relatively less massive body continuously circling around a relatively more massive body within a complete elliptical path because of their gravitational attraction, is the motion of gravity or gravitational orbit.

## Thermodynamics

This section describes properties of matter based on temperature and heat radiation. I will only discuss black body radiation in here; A black body is an imaginary physical body that can continuously absorb any type of energy; A black body also emits energy when it's in thermal equilibrium state; in different temperature this body radiates different type of energy that shows different colors; in room temperature it radiates infrared and appear black in color; as temperature increases its apparent colors change from black to red to yellow to white to blue.

All emitted energy of black body radiation and any radiation are electromagnetic emissions but we mentioned "type of energy" because energy is waves with wavelengths; waves of different wavelengths have different properties. The stars emit electromagnetic emission of different wavelengths; some wavelengths are visible to the human eyes as visible light while some are not but we can observe all of them with wonderful wisdom of scholarly science; almost all astronomical objects emit electromagnetic emissions; even everlastingly emerging; enormously extraordinary end of the visible universe, the Cosmic Microwave Background or CMBR emits electromagnetic emission.

"Excellently extraordinary; enormously eminent Electromagnetic Emissions endlessly encouraging essentially evolving eyes of wonderfully working wisdom of significantly supreme science of highly honorable humans, humanity."

# Electromagnetism

This section of physics describes physical properties; physical interactions between objects with electric charges; causing electromagnetic forces on them; between them; surrounding them.

**Electric Filed**
An Electric Field is an invisible fundamental filed of existing electric charge; exerting electrical effects, electromagnetic force on other objects of charges within this field and also attracting or repelling them.

**Magnetic Field**
A Magnetic field is an invisible fundamental field of magnetic effects; existing elements; affectingly effecting every element with mainly magnetic properties within its surrounding magnetic field.

**Electromagnetic Field**
An Electromagnetic field is an invisible fundamental field of the existing electric and magnetic field of combined magnetic and electric effects, electromagnetic effects; affectingly effecting every existing electromagnetic element within this field. The electromagnetic field is physically produced by any electrically charged objects; this field surrounds this object and affects every electrically charged and magnetic object within it. This field's electromagnetic effects exist everywhere; effects every energy of the universe; the electromagnetic field extend endlessly throughout totally surrounding spacetime.

"Fundamental field from electromagnetic effects endlessly exist everywhere; effects every energy's essential existence and affects everything everywhere entirely; endlessly effects entire endless entirety."

# Electromagnetic Emission; Energy

Every Energy is Electromagnetic Emission or Electromagnetic Radiation; so every energy is waves; different forms of energy are waves of different wavelengths. This radiation consists of electromagnetic waves within electromagnetic filed; these waves are synchronized oscillations of electric and magnetic field. Electromagnetic field is an energy field; it is the combination of electric field and magnetic field; every energy existingly travels throughout surrounding spacetime wonderfully, within this field.

# Light

Photons are the particles of light, energy. The light is electromagnetic radiation; light is energy containing photon particles traveling as waves within the electromagnetic field. Every energy of our universe is electromagnetic emission. Radio waves, microwaves, infrared light, visible light, ultraviolet light, X-rays and gamma rays all are waves of different wavelengths of electromagnetic radiation. Gamma rays have the shortest wavelength in the electromagnetic spectrum and also has highest photon energy. Solar Flares, Pulsars, Quasars, Active Galaxies, collisions of multiple Neutron Stars, collisions of Neutron Stars with a Black Hole and formations of new Black Holes emit Gamma Rays; these are main sources of high energy Gamma Rays.

**Color of Light; Other Objects**
Different colors of light have different wavelengths such as the blue light has a wavelength of 450 - 490 nanometers and the wavelength of red light is 635 – 700 nanometers; blue light has shorter wavelength than red light. Light contains waves; the length between two high points of adjacent waves is its wavelength. You see a particular color of an object because that object absorbs all the wavelengths of light except the wavelength of that particular color of light and then reflect it to your eyes.

# Matter

Matters are elements of the cosmos containing common particles with mass; matters occupy spacetime by having volume. Mass is concentrated energy; everything with mass also has an equivalent amount of energy or in other words, mass and energy are equivalent relative to their mass-energy equivalent physical properties. Matters are made of molecules; molecules consist of Atoms; this is very simple proven physics. There are approximately 37 trillion cells within a human body; one human cell contains approximately a hundred trillion Atoms. The number of Atoms in a single cell is more than the number of stars in the Milky Way, Gigantic Galaxy. One single cell of your body is actually a greatly gigantic galaxy of atoms, relative to Galaxy of stars not by size but by numbers of containing objects.

**Plasma:** There are four states of matter; they are Solid, Liquid, Gas and Plasma. Plasma is an extremely electrically conductive gas; this ionized gas contains negatively charged Electrons and also positively charged Ions. Any area surrounded by Plasma gas is also an extreme electromagnetic field; so it has electromagnetic properties and can conduct electricity. Plasma exists within astronomical Nebulas.

# Mass Energy Equivalence

One of the tremendously triumphant theories of proven physics proved the equivalence of mass-energy that tells that the amount of energy is equal to the amount of mass multiplied by the square of light's constant velocity. This gave humanity the capability to produce enormous energy from significantly small amount of mass, matters and also designed delightful doorways for new seriously surprising science.

# Nuclear Fission

In this process, large atomic nucleus of heavy radioactive elements such as Uranium-235 continuously splits into relatively small nuclei within nuclear reactor and also emits enormous electromagnetic energy. An Atomic Bomb or Nuclear Fission Energy Reactor generate enormous energy using this process.

# Nuclear Fusion

This is the opposite process of Nuclear Fission; the Nuclear Fusion fuse nuclei to generate enormous electromagnetic energy. Relatively low mass nuclei come completely closer; compactly combine or forcefully fuse; form a relatively massive nucleus and also emit enormous electromagnetic energy in this complexly continuous complex process.

This process needs extreme energy to make strong nuclear forces of nuclei to overcome their own electrical repulsive forces to perfectly fuse them into the relatively massive nucleus and also emit extremely enormous electromagnetic energy; so this process requires an extremely high temperature and perfect pressure to constantly continue Nuclear Fusion. Stars are Natural Nuclear Fusion Reactors emitting enormous electromagnetic energies; Main sequence stars, significantly stable Natural Nuclear Reactor relative to other ordinary shining stars.

"Seriously saying; significantly shining Sun's superbly stable "Natural Nuclear Fusion Framework" emitting enormous electromagnetic energy; ensuring extraordinarily elegant Earth's endlessly essential existence of obviously honorable humanity; legitimately living lovely lives longing long lasting love."

# Four Fundamental Forces

There are four fundamental forces of the "Energetically Expanding Endlessly Unbounded Unlimited Universe"; they are described below.

**Strong Nuclear Force**
This is the strongest; short ranged force of cosmos; This fundamental force forcefully holds the nucleus of an atom. This is mostly an attractive force but sometimes repulsive relative to surrounding situations.

**Electromagnetic Force**
This force acts on electrically charged and also magnetic elements to cause electromagnetic effects. This force is long ranged but relatively weaker than the Strong Nuclear Force.

**Weak Nuclear Force**
This short-ranged weak force acts between subatomic particles of atoms to cause radioactive decay and also cause neutrino interactions.

**Gravitational Force**
This is the most long ranged attractive force acting between masses of the Universe within their own ranges. The gravitational force causes orbits of planets around stars and orbits of stars within galaxies.

# Unification of Forces; Electroweak

The two fundamental forces, Electromagnetic and Weak Nuclear Force of four fundamental forces are actually different in normal temperature or low energy environment but they can combine themselves to remain unified; this unification of Electromagnetic and Weak Nuclear force happens when the environmental energy is above Unification Energy, enormously extreme energy typically thermal; in other words, hugely high temperature. The Unification Energy scale is 246 GeV; it's an enormously extreme energy state. After a very short period of the Big Bang, the existing environmental energy was above Unification Energy scale; it was enough to merge Electromagnetic and Weak Nuclear Force to Electroweak Force. This is very important information if you want to understand singularity, mainly the Cosmic Singularity that completely created Cosmos.

# Proven Particle Physics

In this section proven physics of particles and subatomic particles will be explained; Endlessly essential enormously elegant English excellently explaining perfectly proven Particle Physics.

# Standard Model

An Electron is a fundamental particle of the Universe but what about Proton? Proton is not a fundamental particle; it's proven in a great theory called Standard Model; this theory describes the elementary particles of the Unlimited Universe. The standard model also describes three fundamental forces of the Universe; they are Electromagnetic, Strong Nuclear and Weak Nuclear force. The Gravitational Force is actually the curvature of Spacetime explained in the General Relativity section. These forces are caused by their carrier particles; so infact forces also have fundamental particles.

**Elementary Particles**
Elementary Particles are divided into two major groups; they are Elementary Fermions and Elementary Bosons.

**Elementary Fermions:** Quarks and Leptons; they form matter.

**Elementary Bosons:** Gauge Bosons cause force and Scaler Boson cause mass.

**Description of Four Types**
In Standard Model there are four deferent groups of particles; their given names are Quarks, Leptons, Gauge Bosons, Scalar Bosons. The existence of all these elementary particles is proven in hadron colliding experiments; in this experiment, high-velocity particles are collided with each other to split them; then they together form new particles; the results from this entire experiment give new scientific breakthroughs and also prove the existence of elementary particles. Quarks and Leptons are the particles that create matter; they are also called Fermions. The Gauge Bosons are particles that cause the force or carry the force. Finally, the scaler boson causes mass to matters.

# Elementary Particle's Classification

### Quarks
Quarks are elementary particles participating in strong interactions of subatomic particles. There are six types of Quarks and they are Up Quark, Charm Quark, Top Quark, Down Quark, Strange Quark and Bottom Quark.

### Leptons
Leptons are elementary particles participating in Electroweak interactions of subatomic particles. There are six types of Leptons and they are Electron, Muon, Tau, Electron Neutrino, Muon Neutrino, Tau Neutrino.

## Gauge Bosons

Gauge Bosons are the carrier of forces; they act as exchange particles between quarks to cause force. There are four of them; they are Gluon, Photon, Z Boson, W Boson. Photons are electromagnetic force carriers; W and Z Bosons are weak nuclear force carriers. Gluons are the strong nuclear force carriers; there are eight types of Gluons. All the three forces, Electromagnetic Force, Strong Nuclear Force and Weak Nuclear Force are caused by Gauge Bosons; they are the elementary particles for fundamental forces of the Universe.

## Scalar Boson

The last one, most wonderful one, Scalar Boson is the Higgs Boson. The Higgs Boson is an elementary particle of the Universe; Higgs Bosons coexist with the Higgs Field to cause mass to matters; particles of matters. The Higgs Field is a quantum energy field and this field uses the Higgs Boson particle to interact with other particles of matter such as Protons, Electrons, etc. When any matter particle interacts with Higgs Field containing Higgs Bosons, it gains mass; so, the Higgs Bosons are the fundamental particles causing mass to matters. This is the definitional description of mass in Perfectly Proven Physics.

# The Formation of Matter; Matter Making And Atomic Arrangement

Now, let me explain how Particles of Atoms and all Atoms are form. One Up quark and two Down quarks together form a Neutron; two Up quarks and one Down quark together form a Proton; the Gluons work as force carriers to bound the quarks within those subatomic particles with Strong Nuclear force; then the new Neutron and Proton combiningly form an Atomic Nucleus. Electron is an elementary particle; then that newly formed Atomic Nucleus and an Electron together form an entirely new Hydrogen Atom; the electron orbits the nucleus because of the electromagnetic force. This is how different types of Quarks bounded by different types of Gluons form all archetype Atoms; then they together make matter molecules; then these molecules make matters.

**Neutrino**
Neutrinos are similar to electrons but have no charge; these particles can typically pass through most matters; Neutrinos are the most abundant particle in the Unlimited Universe; they travel almost with the speed of light but can never exceed this universal speed limit.

# Antiparticles And Antimatters

The standard model also describes Antiparticle and Antimatter; antiparticles create all antimatter like ordinary particles create all ordinary matters. An antiparticle is actually a particle of exactly same size and same mass as an ordinary particle, but it has opposite charge; so it's called "Antiparticle". An electron with positive charge is an anti-electron; it's also called Positron; A Proton with negative charge is an Antiproton. An Antiproton and an Antielectron together form an Anti-Hydrogen atom, like a proton and an electron forming an ordinary Hydrogen atom. All matters in the Universe have their own associated Antimatters. Photon is not a matter particle but a massless energy particle with no charge; so the Photon is actually its own antiparticle or Photon and Anti-Photon is exactly the same particle. Antimatters are very valuable elements of the Unlimited Universe because they can produce enormous extreme energy; when antimatter collide with ordinary matter they annihilate each other and also generate an enormous amount of extreme energy. Collision of one kg antimatter with one kg normal matter will produce energy of forty-three megatons TNT.

# Particle's Phenomenal Probability; Quantum Mechanics

The Universe of the smaller objects is as wonderful as the larger objects. The particle of light is the photon. The photon is massless but is it a particle only? Light is electromagnetic radiation containing waves. The photon is both particle and wave. In experimental observation like photo-electric effects shows that photon interacts with electrons as a particle and then emits electron. The double slit experiment shows that when light travels through the slits it creates wave patterns on the behind screen; so the light, photon is both particle and wave. Not only photons but even atoms of ordinary matters show wave patterns in the double slit experiment. Atoms and all subatomic particles are both particle and wave; this fact gave birth to quantum mechanics, one of the greatest theory of honorable humanity. Not only every energy but also matter shows perfect properties of waves; wonderful waves wandering within elegantly endless Entirety; everywhen; everywithin; everywhere.

**Wave-Particle Duality; Double Slit Experiment**
The double slit experiment is the most wonderful of all. When you shoot electrons one by one through the two slits they make numerous stripes instead of just two stripes on the detector screen behind the two slits; this pattern of numerous stripes is called wave patterns or the interference pattern. The double slit experiment for electrons or other particles gives the same result as the double slit experiment of light, described before. Light and all other particles have wave properties; this is called the wave-particle duality of energy and matter.

These waves are not just waves of electrons like the waves of water but instead, they are probability waves; the size of a wave in a particular position actually tells you the possibility of finding electrons there. Higher size of wave gives you a higher possibility of finding electrons on that particular position. When an electron is shot through the double slit you can never tell where it will land on the behind screen but when enough electrons are shot you can find the locations of different percentages of electrons on different stripes by calculating waves; to know that you will need to calculate the probability wave of those electrons using Schrödinger equation. This was proofed again and again in experiments so the world of particles is also a world of probability, Quantum Mechanics, unlimited uncertainty.

**Uncertainty Principle**
The uncertainty principle states that if the position of a quantum particle or quanta is measurable then its momentum is unmeasurable; if its momentum is measurable then its position is unmeasurable; so measuring the position and momentum of the same quantum particle is impossible. This marvelous mystery makes the quantum world of particles very mysterious; because of the uncertainty principle any quantum particle can pop in and out from existence; it can exist in every position of that system as waves. Any unobserved particle endlessly exists in indeterminable infinite positions but when observed, the particle's positional uncertainty disappears; then that particle appears to the observer in its particular position.

The observable universe is also made of particles; so it's also a world of uncertainty; not certainty. The intelligent human observers give meaning to the existence of unlimited universe, relative to human observation and understanding; this is true in infinite ways. Without Wise Cosmic Observer; constantly continuous Cosmos cannot endlessly exist. Understanding; calculating; predicting the movements, behaviors, properties of Quanta, quantum particles by completely calculating probabilistic possibilities is Quantum Mechanics.

Electronics is directly related to Electrons. The marvelously magnificent modern humanity invented all the modern electronic devices after understanding Quantum Mechanics; no electronic device is possible without understanding the behaviors and movements of Electrons within macrocosmic matters and also vacant vacuum.

## Quantum Entanglement
The quantum entanglement is that, if you have a pair of entangled quantum particles and change the quantum properties of one particle then it affects the other paired particle. These physical phenomena instantly happen, even when the distance between paired particles is infinite; it's like a universal bond between quantum entangled, paired particles that doesn't depend on time. In Quantum Entanglement, two particles can become entangled if they are near to one another; then their physical property pairs with one another. These particles can remain entangled to one another, even when the distance between them is changed to infinity.

Like all the particles, an Electron spins randomly; so you can never tell its direction of spinning without measuring it, but when you measure it; you will find that an electron either spinning Clockwise or Anti-Clockwise in its own Axis. The Clockwise spin is also called Spin Up and Anti-Clockwise spin is called Spin Down. When two Entangled Electrons are separated by infinite distance and you measured the spin of one Electron and found that it's spinning Clock Wise; then the another Electron is surely spinning Anti-Clockwise. The astonishingly amazing futuristic fact is that, if you change the spin or properties of one of the electron then the other one also changes instantly without depending on Spacetime. "That's a fantastically futuristic fact because "because of Quantum Entanglement perfectly paired particles can completely teleported to distantly different spanningly surrounding SpaceTime"; The Time: Previous Past; Passing Present; Further Future; The Tremendously Terrific Time Travel; The Time Traveler Travelling Triumphantly."

# Empty Expanse; Endless Energy of Vacant Space

Particles exist in the Quantum World, a world that cannot be observed with human eyes but can be understandingly explained with human wisdom. The world of Quantum particles is marvelously mysterious; within this world, particular particles known as virtual particles appear and vanish and then reappear again on every region of Spacetime.

Virtual Particle Pairs continuously exist in every region of expanding Spacetime of the Unlimited Universe; each of these particle pairs continuously contains one normal virtual particle and another anti-virtual particle; these two particles are constantly created and destroyed; continuously happening phenomena. These phenomena are outstanding observable facts in every region of empty space. Even when space seems entirely empty, it's not entirely empty because these pairs of virtual particles pop in and out in every empty space. One normal virtual particle and one anti-virtual particle created from nothing exist in every empty space; then they collide to destroy themselves; then reappear again; then destroy themselves by colliding again; these constant creations and deathless destructions are always happening in every entirely empty region of the Universe.

## Quantum Fluctuation

Vacuum Fluctuation is also known as Quantum fluctuation; it is the measurable energy change in any single point of empty space. Because of the uncertainty principle of quantum mechanics, this Quantum Fluctuations continuously happens in every point of empty space; this also causes the creation of virtual particle pairs. When any normal particle and antiparticle annihilate themselves in collisions they emit electromagnetic energy; so the destruction of virtual particles also emit energy. This information is also very important to understand the evolution of the Universe.

Ashraf A.

# Atoms; Electron's Orbit

Inside an Atom, an Atomic Nucleus consisting of neutron and proton exists; its Electrons orbit that nucleus. Atomic nucleus exists in the center of an Atom; it's made of compactly combined Neutrons and Protons. The Electrons orbit an Atomic Nucleus because of electromagnetic force. This orbit is not like "planets orbiting the Sun" but a wavelike orbit; because of the uncertainty principle, this orbit looks like an Electron Cloud. An Electron is also both particle and wave; so it cannot have a particular shape. When Electron interacts with other particles its shape is like a small spherical point but when it's inside of an Atom, the "Shape of Electrons" varies on different Energy States. When the Energy State changes the shape of the Electrons changes constantly. The Electrons orbit the nucleus of an atom because of charge differences; when an Electron changes its orbital position because of receiving and emitting energy it just vanishes from one orbit; then appears on the another. Electrons exist in fixed or dedicated orbits inside an Atom within Electron Cloud. When an atom is heated or given energy its electron leaps to an upward orbit; when an electron leaps to a downward orbit it emits energy of specific wavelength or color of light; this vanishing of electrons from one orbit and then reappearing on other orbit is called leaps of electrons, Quantum Leap; this quantum leap happens because the energy of electrons comes as constant quantity that cannot be subdivided; this minimum constant quantity of energy is called Quanta. The energy of an Electron is quantized; from this, this theory was named Quantum Mechanics.

"Photons, particles of Light, Energy; every energy; every electron; every elementary particle; particles of atom; all particles can constantly exhibit emerging wavelike behaviors because of wonderful wavelike properties; so they also move like waves; understanding; calculating; predating their wonderful wavelike marvelous movement is Quantum Mechanics."

# Electron Degeneracy Pressure

In the same system or volume, two Electrons with same spins cannot continue their existence with same exact energy state. If a group of Electrons with same spin fully fill a lower energy level or say level one, the other group fills the higher level or level two; as Electrons increase in number, they fill higher and higher energy levels. Higher energy level Electrons also have higher energies; so they move more fast and these moving Electrons cause powerful pressure or Electron Degeneracy Pressure; this pressure supports stars and star cores to prevent gravitational collapse.

# Grandly Governing Gravity; General Relativity

General Relativity is a theory of Spacetime Curvature, that defines Gravity as a spacetime curvature caused by Mass.

## Simplifying Spacetime

There are four Dimensions in the Universe; they are three Spatial Dimensions and one Time Dimension. Everything in the universe exists within these Dimensions; all these four Dimensions are complexly combined together; they cannot be separated, so this cosmic combination of four dimensions is Spacetime. Every point of empty or not empty space is Spacetime; everything in the universe moves within Spacetime. If you consider Spacetime as a thing than it is the most abundant thing in the universe. Spacetime doesn't only exist in entirely empty regions but it also exists within every matter; a hydrogen atom consists of 99.9999999999996 percent empty space or Spacetime. Even the human body is almost 99.999 percent empty space or Spacetime. Not only human body but everything you observe is almost 99.999 percent empty space. So, never remain upset when you feel empty inside, in fact, you are entirely empty inside.

# Curvature of Spacetime

Spacetime is like flexible fabric; it can be curved; it can be twisted; it can also be warped; this is what mass does to Spacetime; every existing mass curves Spacetime. Because of the Spacetime curvature caused by a massive object, every other relatively less massive object existing around it always falls toward the center of that relatively more massive object; this constantly continuous essential effect is called Gravity. The Spacetime curvature is mathematical in human understanding but it has observable effects. This curvature can be imagined as the effect of placing a seven kg iron on the center of a trampoline; that seven kg iron on the trampoline will cause a downward curvature in the fabric of that trampoline; then if you place some tiny balls of relatively smaller masses on that curvature they will always fall toward that seven kg iron. Everything above the Earth is always falling toward its center but we observe it as attractions of gravitation; even everything on the surface of the Earth is also falling but the surface is providing enough upward force to keep it stably stationary.

# Gravitational Time Dilation

The Time-Dimension coexists with three spatial dimensions and it cannot be separated. What happens to the time dimension in curved Spacetime? In this case, Time Dilation happens; Every massive object slows down Time relative to the Time of a distant uncurved region of the Universe, where no mass is present.

The more you move toward any massive object causing gravity, the more it slows down time surrounding you. In a building with hundred floors, the Time in its ground floor is slower relative to the hundredth floor but you cannot notice it because that time dilation is very small. A massive Black Hole can cause extreme time dilation; if you can survive near a massive Black Hole, every second surrounding you will be equal to thousands of Earth's years relative to Earth's Time.

# Gravitational Lensing

Spacetime can be curved by a massive object, so when light travels near that massive object it travels through that curved Spacetime and as a result the path of that light within Spacetime also curves relative to that curvature. This effect can create Gravitational Lensing of Light around any massive body. Galaxies generate Gravitational Lensing around it; Black Holes also cause Gravitational Lensing around it. A massive Black Hole can cause a temporary unstable orbit of photon or light around it; so the Gravitational Lensing and Orbit of Photons around a Black Hole can be observed as a round ring of circularly luminous light.

# Gravitational Wave

Spacetime also has waves; it means that waves can be created in Spacetime because of its flexibility. When two or more massive objects collide with each other they create Gravitational Waves; these are waves in the fabric of Spacetime. The collision of Neutron Stars causes Gravitational Waves; The merging process of two or more massive Black Holes also creates Great Gravitational Waves. Gravitational Waves always travel at the speed of light; so the speed of Gravity is same as the speed of light.

These all are observable facts of the Ultimately Unlimited Universe.

# Time

Time is the most mysterious thing in the Unbound Universe; it is a dimension like the other three spatial dimensions. Time does not move but energies and matters move within the dimension of Time. Because of the motion of everything in the Universe and the accelerating expansion of the Universe, every energy and matter is in constantly continuous motion within Time Dimension; so the motions of the Universe on the Spacetime cannot be stopped; the endless perception of time will always constantly move on toward the forward forthcoming future.

# Understand Units

Units are standard quantities to meaningfully measure magnitude. In the imaginary spatial dimension system of three dimensions, divide each dimension or imaginary linear line into ten divisions; each division has a length of one meter, so every single dimension has a total length of ten meters.

Using these imaginary dimensions of the spatial system, you can definingly describe; meaningfully measure any three-dimensional objects with length, depth or X-dimension, width or Y-dimension, height or Z-dimension. If the total length of X, Y, Z of your imaginary dimension system is only ten meters then you can only measure objects with size less or equal to ten meters; for larger objects, you will need a larger imaginary dimension system but don't worry you can increase it infinitely. You can create dimension system with any unit of Length, Time, Mass, Energy or other Units to define; describe different diagrams of meaningful measurements of macrocosmic magnitudes.

Common units of physics are:

**Unit of Time:** Second;
**Unit of Mass:** Kilogram;
**Unit of distance or Length:** Meter.
Unit of distance greater than Meter is **Kilometer**;
one Kilometer is equal to one thousand Meters.

**Unit of Energy:** Joule.
**Unit of Luminosity:** This is Joule per second; Watt.
**Unit of Temperature:** Kelvin; Fahrenheit; Celsius.

# Power of Ten

The total number of digits in a large numerical value can be obtained easily by understanding the power of ten; This "POWER" of "TEN" is actually a "NUMBER"; this "NUMBER" expresses the number of digits in a greatly large or very small number with many digits. Understanding "Power of Ten" gives you the capability to realize the largeness or smallness of numerical values expressed using "Power of Ten", relative to other ordinary numerical values; completely counting the "Digit Number" of any numerical value expressed using "Power of Ten", is enough to imagine its largeness or smallness; it will be very helpful to understand greatly large or small scales of the Universe. This is the only mathematical section.

**Digit Number of Large Numerical Values expressed using Positive Power of Ten**
"Total Digit Number" in "Positive Power of Ten" is (1 + "Value of Power on Ten")

**Digit Number of Small Numerical Values expressed using Negative Power of Ten**
"Total Digit Number" in "Negative Power of Ten" is ("Number of Digits, except Power of Ten part" + "Value of Power on Ten").

# Positive Power of Ten

Power of Ten is a special type of number expressing system, used to express numbers with too many digits; using this system any number with any amount of digits can be expressed. This "power" is any number "n" that expresses the number of multiplications of "Ten"; if n equal to two, then power of ten is $10^{(n=2)}$ or $10^2$ which is equal to "10 multiplied by 10" or 100; $10^3$ equal to $(10 \times 10 \times 10)$ equal to 1000; so simply saying the Power(n) of Ten is the number of digits after the first digit or the Total Digit Number of any number expressed using "Power of Ten" is actually (1+n), such as the number of digit in $10^3$ is (1+3) equal to 4 or (1, 0, 0, 0). Now, by multiplying the "Power of Ten part or $(10^n)$" with a significant small number, any large number with many digits can be excellently expressed. Let's give some example for better understanding, 1 multiplied by $10^7$ is equal to 10,000,000; 1.23 multiplied by $10^7$ is equal to 12,300,000; 3 multiplied by $10^7$ is equal to 30,000,000; 7.777 multiplied by $10^{30}$ equal to 7,777,000,000,000,000,000,000,000,000,000; these are actually expressed as $7.777 \times 10^{30}$, a number of thirty-one digits.

# Negative Power of Ten

The "Negative Power of Ten" or 10^(-n) expresses very small numerical values with mainly many "Zeros" after the decimal point " 0. " such as "0.00000"; to keep this simple just remember that the "Number of Zeros, before the first appearance of any other numeral except zero" in a small numerical value expressed using "Negative Power of Ten" is "n" and the total number of digits in this case is (n + number of digits except "Negative Power of Ten" part); for example 1 multiplied by 10^-7 is equal to 0.0000001; 1.23 multiplied by 10^-7 is equal to 0.000000123; 1.101 multiplied by 10^7 is equal to 0.0000001101; 7.777 multiplied by 10^-30 equal to 0.000000000000000000000000007777.

These are actually expressed as $7.777 \times 10^{-30}$, a very small thirty-four digit number; so the total number of digits is simply (n + "Digit Number" in "Numeral Value" before "Multiplication or ×"). If you can understand the number of digits of Power of Ten, it's enough to go further.

# Common Cosmic Calculation; Cosmic Units

These units are for measuring cosmic scale Size; Mass; Energy; Luminosity. Planets, Stars, Black Holes, Galaxies or scales beyond these cosmic objects are meaningfully measured and commonly calculated using Colossal Cosmic Units.

# Planetary Scale

For measuring mass, size, radius of planetary scale ordinary objects:

**For measuring Size or Radius:** Earth Radius or $R_{\oplus}$; the radius of the Earth.
One Earth Radius is equal to 6,371 kilometers.

**For measuring Mass:** Earth Mass or $M_{\oplus}$, the mass of the Earth.
One Earth Mass is equal to $5.972 \times 10^{24}$ kilograms, a twenty-five digit number.

# Stellar Scale or Greater

For measuring cosmic objects beyond Planetary Scale or Stellar Scale:

**For measuring Size or Radius:** Solar Radius or $R_{\odot}$, the radius of the Sun.
One Solar Radius is equal to 695,510 kilometers.

**For measuring Mass:** Solar Mass or $M_{\odot}$, the mass of the Sun.
One Solar Mass is equal to $1.989 \times 10^{30}$ kilograms, a thirty-one digit number.

**For measuring Luminosity:** Solar Luminosity or $L_{\odot}$, the luminosity of the Sun.
One Solar Luminosity is equal to $3.828 \times 10^{26}$ watts, a twenty-seven digit number.

# Cosmic Scale of Length

For measuring greatly large Cosmic Length or Distance:

**Light Second:** the distance light travels in one second; it is equal to
299,792 km, obtained from the constant velocity of light 2.998 $\times 10^8$ m/s.
**Light Minute:** the distance light travels in one minute; it is 1.799 $\times 10^7$ Km.

**Light Years or** *Ly:* the distance light travels in one year; it is equal to 9.461 $\times 10^{12}$ or 9.461 trillion kilometers; it's a thirteen digit number.

**Parsec or** *Pc:* Parsec is a larger unit than a Light Year; one Parsec is equal to approximately 3.26 Light Years or 3.086 $\times 10^{13}$ km.

**Kiloparsec or** *Kpc:* A Kiloparsec is larger than a Parsec; One Kiloparsec is equal to 1,000 Parsecs or 3.086 $\times 10^{16}$ km.

**Megaparsec or** *Mpc:* A Megaparsec is larger than a Kiloparsec; One Megaparsec is equal to 1,000 Kiloparsecs or 3.086 $\times 10^{19}$ km.

**Gigaparsec or** *Gpc:* This is the largest unit of length; One Gigaparsec is equal to 1,000 Megaparsecs. The Observable Universe is approximately 28.5 Gigaparsecs.

Ashraf A.

# Considerable Cosmic Circumstances Perfectly Profound Phenomena

In here, we will wonderfully discuss different delightful Cosmic Circumstances, Phenomena; these are always happening within Cosmos; Creations, Formations, Evolutions of astronomical objects are also parts of this section.

## Marvelous Macrocosmic Motions

The Unlimited Universe is in constant Motion. All Atoms to shining stars to gigantic galaxies; every existing entity within the Universe is always moving in SpaceTime. Marvelous motions mostly making magnificent macrocosms meaningful.

## Gravitational Motion

Almost all the motion of the Universe is gravitational motion but we will consider every kind of opposite directional motion, moving opposite to the attraction gravity as non gravitational motion.

### Axial Rotation

Axial Rotation is the rotational motion of objects on its own axis because of its own gravity; attracting its every particle toward the center. This motion is happening everywhere; the Earth is continuously rotating on its own axis; the moon is also rotating on its own axis; even the Sun is also rotating on its own axis.

### Revolution, Orbital Motion
The rotation of relatively less massive bodies around relatively more massive bodies is revolution or orbital motion. Orbital motion, most marvelously materializing motion, mostly emerging everywhere in the Universe; the Moon is orbiting the Earth; the Earth is orbiting the Sun; the Sun is orbiting the galactic center; All the stars within all the galaxies orbit their own galactic centers.

### Galactic Motion
This is actually the orbital motion of galactic objects within galaxies caused by galactic gravitation; because of this stable uniform motion, all the galaxies actually appear roundly rotating its own central axis or central core.

# Gravitational Motion, Non Orbital

Objects moving toward each other because of gravity but not stably orbiting each other is non orbital motion; this type of motions mostly results collision, consumption and merging of objects.

# Cataclysmic Collision

This is always happening in the cosmos but stable orbital motions are greatly high in number relative to collisions; that is why most of the cosmic objects are still existing in perfect shapes; conditions; motions.

## Asteroid Collisions

Asteroids compactly exist within Asteroid Belts; so all the Asteroids within their particular Asteroid Belt continue their own orbital motions while remaining compactly close to each other; so Asteroids are always colliding with each other within all Asteroid belts of the Unlimited Universe. Sometimes Asteroid Collisions can cause Asteroid ejections; throw Asteroids out of their particular Asteroid Belt region; then those ejected Asteroids can collide with planets, moon or other objects.

## Planetary Collision

When two or more massive bodies within a solar system, such as planets are orbiting their specific star while remaining near to each other; they can come closer to each other to cause cataclysmic collision, because of gravity; these planetary collisions can cause complete cosmic creations like natural satellites or asteroids.

## Stellar Collisions

When two or more star systems are orbiting the galactic center while remaining near to each other; they can come closer to each other because of their gravitational attraction; then they can collide with each other; these collisions can cause planetary collisions and Stellar collisions, the collisions of two or more stars. Multiple stars orbiting each other in Binary, Trinary type star system can cause collision too, if they come completely close to each other because of gravitation or one of the massive star grow large in size to cross other star's orbital path.

**Galactic Center; Continuous Cataclysmic Collisions**
Within any galactic central core, massive objects such as massive stars, asteroids, nebulas, black holes are in constantly continuous compact orbit around their own galactic center; so countless cataclysmic collisions are frequent near the galactic center because of star density or galactic object density in the galactic core. Mostly massive stars are continuously colliding with each other near the galactic center; causing cataclysmic circumstances; creating excessively extreme environment everywhere; everywhen; everywhiting.

# Complete Consumption

### Star Consuming Planets
Stars can grow very large in size when going through Stellar Evolution process; so the size of a growing star can also extend beyond its inner solar system boundary; when this happens that large star can collidingly consume every object or orbiting planet within its inner solar system. This is actually a collision but a star is greatly massive relative to planets; so when a star consumes a planet it appears as it is consuming it.

### Black Hole Consuming Cosmic Objects
The only true consumption is, a Black Hole consuming other objects; when other objects such as stars come close to a black hole and also cross its inescapable gravitational boundary because of that Black Hole's extreme gravity; it tears those shining stars to form accretion disk, a disk of complexly compact circularly moving highly hot particles of star stuffs; then it continuously consumes those particles.

Black Holes formed from Supernovae also use the same process to consume surrounding star stuffs or other objects. Any active ascertain-disk also emit enormous electromagnetic emission to vast distances; so this cosmic object is one of the largely luminous outstanding objects of the "Ultimately Unfixed Unboundedly Unlimited Universe, Endlessly Expansive Eminently Enormous Entirety."

# Marvelous Merging

### Solar System Merging
Solar system collisions can merge multiple solar systems together to form a more massive solar system; this merging can cause planetary collisions or Stellar collisions rarely.

### Black Hole Merging
The most marvelous merging is the merging of Black Holes; Black holes can grow more massive by constantly consuming other ordinary orbiting objects and also by merging with other Black Holes; two or more massive Black Holes orbiting each other can continuously reduce their own orbital positional distance to come completely close to each other because of their gravity; then merge with one another to become a more massive Black Hole. Black Holes are greatly massive objects with extreme gravity that they can cause remarkable ripples in Spacetime; so the process of their merging causes gravitational waves traveling at the speed of light across vast distances of the cosmos.

"Magnificently merging Black Hole's compactly close complete circular courses continuously creating cataclysmic circumstances; causing reoccurring reverberations; remarkably rippling flexible fabric of surrounding SpaceTime to constantly cause great Gravitational Waves, wonderful waves of surrounding SpaceTime."

## Galaxy Merging

Massive Galaxies bounded by their gravity can collidingly merge to form a more massive gigantic Galaxy. Our home galaxy Milky Way and our neighbor galaxy Andromeda are actually in a collision course; they are continuously approaching toward each other; so they will completely collide; merge marvelously in forthcoming future; fabulously form an abundantly more massive greatly gigantic galaxy, more massive Milky Meda after approximately four billion Earth years. This magnificent merging of massive galaxies can cause very few or no stellar collisions because of vast distances between stars within galaxies but it will obviously merge their central Super Massive Black Holes to form a new central Hyper Massive Black Hole. The newly formed Milky Meda galaxy will have all the stars of Milky Way and also Andromeda inside its diameter; those stars will also orbit its newly formed Hyper Massive Black Hole at the center; this is how gigantic galaxies extraordinarily evolve.

# Non Gravitational Motion

Motions of objects opposite the direction of gravitation are considered as non gravitational motions.

## Ejection of Planet

Planetary objects can escape their solar systems because of explosions of stars or collisions caused by other objects near them. These rare phenomena of ejecting planetary objects outside origin solar systems into interstellar medium are called ejection of planets; these ejected planets freely move outside its own solar system in the interstellar medium.

## Ejection of Star

Stars usually remain in orbital motion within the gravitational boundary of its origin Galaxy but sometimes supernova explosions of relatively more massive stars near it can cause those small stars to forcefully escape the galactic gravitational boundary; then those special stars are ejected into intergalactic medium to freely move outside their galaxies. These rare phenomena perfectly happen in multiple star system such as trinary, binary because of the extreme explosion of the most massive central star relative to less massive orbiting stars.

## Dark Energy; Entirety's Endless Escalating Expansion

The only original non gravitational motion is the accelerating expansion of the Universe; causing completely by Dark Energy; this motion is directly divergent; dedicatedly dominating greatly governing Gravitation; this Energy entirely exists everywhere; supremely spreading surrounding space of entire Universe. Is this a motion? Yes, it can be considered as the motion of surrounding space but not ordinary objects; Dark Energy is the energy of empty space that causing accelerating expansion of spacetime. The spacetime between gravitationally separated objects or groups of objects is expanding rapidly; this gravitational separation denotes different galaxy groups, unaffectedly unbounded by their governing gravitation; gigantic galaxy groups not sharing same gravitationally governing boundary because of vast distances between them. Galaxies near to our Milky Way are gravitationally bounded and so they are moving toward each other or us, but distant galaxies are not gravitationally bounded to Milky Way; so they are moving further away because of accelerating expansion of spacetime caused by Dark Energy; this acceleration of expanding spacetime significantly increases relative to distance; so further galaxies are moving away more faster; most further galaxies are moving faster than the speed of light relative to the Milky Way.

# Electromagnetic Energy Emissions

### Star's Light Emission
This is the process of energy radiation by Stars. All the stars emit photons, particles of light; those particles always remain in constant motion, move at the constant velocity of light across cosmos.

### Explosion of Stars; Supernova; Hypernova
When a massive star's gravity cannot continue its fusion process anymore, it crushingly; compressingly causes the death of that star by creating cataclysmically extreme explosion known as Supernova and Hypernova. This type of explosions emits enormous electromagnetic energy across vast distances; so these explosions are very luminous. Powerful ground and space telescope can observe these luminous explosions; the accelerating expansion of the Universe is proven by observing these largely luminous extreme explosions at different distant regions of the Universe.

### Pulsar's Energy Emission
Pulsar, a pulsating neutron star pulsingly emit enormous electromagnetic emissions. When these pulsating emissions occur toward the direction of Earth, profoundly powerful telescopes can correctly detect it.

### Quasar's Energy Emission
Quasars are active galactic cores; they are also great sources of enormous electromagnetic emissions. These extremely luminous astronomical objects emit enormous extreme electromagnetic energy across great distances of the Universe.

## Radio Emission

Many types of astronomical objects of the universe emit radio emission, electromagnetic energy of different wavelengths known as radio wave; powerful radio telescopes can detect them. The Sun, Jupiter, Supernova Remnant, Neutron stars, Pulsars and Star forming regions within Nebulas are significant sources of radio emission. Some galaxies known as Radio Galaxy also emit strong radio emission. The galactic center of milky way containing the Super Massive Black Hole is also a significant source of powerful radio emission. The most profoundly powerful source of radio emission is a quasar; it also emits enormous electromagnetic energies of different wavelengths. The Cosmic Microwave Background (CMBR) endlessly emits radio emission everywhere.

## Gamma Ray Emission

Gamma rays have the shortest wavelength of only $10^{-12}$ meters; As the wavelength of electromagnetic emission decreases, its energy increases relatively; so Gamma Rays always have highest energy of any electromagnetic emission. Solar Flares, Cosmic Rays, Magnetars, Pulsars and Quasars are strong sources of Gamma Rays. Collision of multiple Neutron stars, Neutron stars with Black Holes and active Black Holes are also great sources of Gamma Rays. Superluminous Supernovae, significantly strong sources of enormous extreme energy Gamma Rays.

## Microwave Emission

Different types of astronomical sources emit different types of electromagnetic emissions; observing electromagnetic emission of different wavelengths is the wonderful way to outstandingly observe; undoubtedly understand the marvelously mysterious macrocosms; mesmerizing meanings.

The most distant source of electromagnetic radiation is Cosmic Microwave Background Radiation radiating microwave, electromagnetic energy.

### Neutrino Emission

Neutrinos are elementary particles with no electric charge; stars fusion cores, Supernovae and Hypernovae, Supernova Remnants are astronomical sources of Neutrino emissions. The tremendous Big Bang emitted enormous emissions; ejected electrically neutral Neutrinos.

# Remarkable Reflections

### Visible Light Reflected by Moons

These are common phenomena of natural satellites reflecting the light of star; the Moon of the Earth reflects the light of the Sun toward the Earth and we observe it as a glowing moon.

### Visible Light Reflected by Planets

Planets also reflect light of their stars to show their colors; these phenomena can be observed from outer space.

### Visible Light Reflected by Nebulas

Nebulas reflect light of their nearby stars; they appear as colorfully glowing colossal clouds of different random shapes, because of light reflection.

# Luminous Light

### Visible Lights of Galaxies
Galaxies are mainly made of colorfully shining stars; those stars emit enormous electromagnetic energy including visible lights; so galaxies appear as colorfully glowing gorgeously gigantic structure of luminous lights; less dense galactic regions glow less and relatively denser regions glow more; these different divisions of glowing lights of any galaxy; together create a gorgeously glowing galactic effect.

### Visible Light of Galactic Gravitational Ring
This is the effect of gravitation known as gravitational lensing; massive objects with extreme gravity cause this effect; massive galaxies bend light's path to cause a vastly visible round ring of glowing light around them.

### Visible Light of Black Hole's Gravitational Ring
Black Holes are massive bodies with extreme gravity; so Black Holes also cause a vastly visible round ring of glowing light around them. These roundly rings of gorgeously glowing gravitational light lensing are amazingly artistic and also awesomely astonishing to outstandingly observe.

# Complete Cosmic Construction; Fascinatingly Flawless Formations

In here, we will discuss the formations and evolutions of different cosmic objects.

## Accretion Process

Accretion Process is collectingly combining proximate particles to grow greatly massive momentarily; most cosmic objects such as stars and planets are formed by Accretion processes. In Accretion process, gaseous matters within its accretion disk spin very fast and attract more matters around it because of its gravity, to become more massive time to time; then turn themselves by centrally concentrating into most massive macrocosmic objects, relative to its starting source; state; shape; size.

## Interstellar Cloud

Nebulas are actually interstellar clouds; these clouds mainly contain normal gas of elements, plasma and dusts; their density of containing elements relative to their size differ them from surrounding interstellar medium or ordinary outer space.

## Molecular Cloud

Molecular Clouds are special type of interstellar clouds but these clouds are very special because new Stars form within them. Molecular clouds have perfect properties; perfect pressure; perfect density relative to their sizes to form new molecules, mainly molecular Hydrogens. Within Molecular Clouds exist more and less dense regions; highly denser regions called clumps initially start star formation process; when a clump's gravity is enough to forcefully collapse its elements; it initially begins star formation.

### Giant Molecular Cloud

These are the most massive and large sized molecular clouds; the mass of these clouds ranges from $10^3$ to $10^7$ solar masses and size ranges from 15 to 600 light years in diameter.

### Small Molecular Cloud

These are least massive small sized molecular clouds; their mass ranges from a hundred to less than thousand solar masses.

# Star System Formation; Evolution

Star systems form from significantly small sections of large Molecular Clouds; these portions of clouds completely collapse because of gravity to start accretion disk processes; then those rapidly spinning disks form stars in their center; then the other left over elements form planets moons and other ordinary objects. Our own solar system also formed from a sixty-five light years sized Molecular Cloud; that cloud contained mainly hydrogen with relatively less amount of helium. There are three steps of solar system formation and evolution; they are described below.

### Initial, Pre-Solar Nebula

This is the initial state of a significantly small section of a large Molecular Cloud; this relatively dense gaseous small section will gravitationally collapse to concentrate most of its mass toward the center to start the formation of a new star or Protostar. In this initial phase of star system formation, the Pre-Solar Nebula's extreme gravitational attraction toward the center is attractingly moving most matters toward the center and also causing it to rotate rapidly; this rotation is continuously changing solar nebula's initial shape to a disk-like shape.

## Mid-Stage, Proto-Star Formation
In this middle phase of star system formation, the Protostar gained its initial low mass state and still attracting more matters into itself to become more massive perfect Protostar; the surrounding gaseous elements are still rotating very fast and almost changed itself to a perfectly round disk shape.

## Protoplanetary Disk
In this phase of star system formation, the protostar is fully formed and the shape of its surrounding is now completely disk-like, a circumstellar disk that remarkably rotating rapidly. The newly formed dense hot Protostar is not massive enough to star hydrogen fusion; so it is still consuming mass to become a shining star capable of fusing hydrogen to emit light. Accretion processes are happening in many regions within the area of Protoplanetary Disk in order to form planets and planetesimals.

Time to time, that perfect Protostar will gain more and more mass because of its gravity; then it will become a main sequence star to fuse hydrogen and emit electromagnetic energy, light; the leftover elements in the surrounding will form planets by accretion process; then those planets will eventually cool down to take their final form by capturing relatively small masses and bodies around them. Finally, the fully formed shining star's solar flares and radiations will completely clear dusts and other small particles within its system to form the Heliosphere.

# Standard Structural Stratification of Greatly Gigantic Galaxies

A galaxy consists of mainly, many star systems containing shining Stars. Almost All gigantic galaxies contain Super Massive Black Holes in their central cores. All the star systems with stars gravitationally orbit within their galaxies, around the galactic central cores with relatively similar velocities. The simple structure of galaxies is a galactic central core with a Super Massive Black Hole; roundly surrounded by enormous amount of stars; shining stars surrounding structures mainly depends on galactic type. The entire Universe consists of mainly, many glowing galaxies containing shining stars; so significantly simple structure of the Universe is mainly, many multiple gigantic groups of gravitationally grouped gigantic galaxies; gravitationally attracted and adjoined wonderfully with or onward one another and then totally together.

**Structural Stratifications of Galaxies**
There are three common classes of galaxies; they are Spiral Galaxy, Lenticular Galaxy, Elliptical Galaxy.

**Spiral Galaxy**
This type of galaxy has a flatten disk shaped structure; so, surrounding shining stars within those galaxies combinedly create completely flatten disk shaped surrounding structure; significantly cause complex flatten disk shape structure to those gigantic galaxies. This type of galaxies also has centrally connected several spiral arms and shining stars beautifully born within those shining spiral arms. Spiral galaxies are also classified based on their significant structure into three types; they are Normal Spirals, Barred Spirals, Intermediate Spirals.

## Spirals or Normal Spirals

These galaxies have commonly two or specially several spiral arms as pinwheels with central bulge; those spiral arms are connected to the central bulge; the central bulge is a circular core or spherical core at the galactic center containing tightly packed groups of stars, supermassive stars; specially several significant supermassive Black Holes or ordinarily only one in normal cases. Spiral galaxies are commonly noted using "S"; then the letters "a to c" are used after "S" to represent the tightness of spiral arms and also the size of galactic cores; as letters change from a to c, spiral arms tightness and central core size decree relatively. In extended classification letter "A" is used after "S" to represent "no central Bar" and after that, the letters "a to d" are used to represent tightness of the spiral arms or normality of shape of the spiral arms and letter "m" is used to represent the irregular shape. The letters "a to d" represent "more tightness of the spiral arms" to "less tightness of spiral arms" toward the galactic center or "normal shaped spiral arms" to "diffused shaped spiral arms"; as the letters change from a to d, the galactic spiral arms extend outward from the central core. All classifications are also described below.

**Sa or SAa:** Have a large bright bulge with very tightly joined spiral arms; compactly concentrated toward the central bulge; the spiral arms are inwardly close to the center and as type changes the spiral arms extend more outward relatively.

**Sb or SAb:** Have a normal bright bulge with normal tightly joined spiral arms; those spiral arms are extended outward; relative to the previous type.

**Sc or SAc:** Have a smaller bright bulge with less tightly joined spiral arms; those spiral arms are extended more outward; relative to previous types.

**SAd:** Have a smallest bright bulge with least tightly joined spiral arms; those spiral arms are greatly extended outward; relative to previous types.

**SAm:** Have a central bulge but the surrounding structures are randomly irregular.

## Barred Spiral

This type of galaxies has a bright barred central core or bar-shaped central core surrounded by commonly two spiral arms or specially several spiral arms. This type is denoted using "SB" and the other classification cases are same like the previous type or "Spiral". The Milky Way our home galaxy is a barred spiral galaxy. This type's classification is described below.

**SBa:** Have a large bright bar at the center with very tightly joined spiral arms; compactly concentrated toward the center; those spiral arms are inwardly close to the center and as the type changes the spiral arms extend more outward relatively.

**SBb:** Have a central bright bar with normal tightly joined spiral arms; those spiral arms are extended outward; relative to the previous type.

**SBc:** Have a central bright bar with less tightly joined spiral arms; those spiral arms are extended more outward; relative to previous types.

**SBd:** Have a central bright bar with least tightly joined spiral arms; those spiral arms are greatly extended outward; relative to previous types.

**SBm:** Have a small central bar but the surrounding structures are randomly irregular.

## Intermediate Spirals
This rare type of special galaxies shows both the feature of Spiral and Barred Spiral Galaxies; denoted with "SAB" and also classified as described above.

## Lenticular Galaxy
This type of galaxies has flattened disk shaped structure with complex central bulge but has no visible spiral arms and denoted with "S0". Any galaxy with a flat disk shape structure with no spiral arm is considered as Lenticular Galaxy.

## Elliptical Galaxy
This type of galaxies has ellipsoidal shaped structure or elliptically spherical structure; so they appear as elliptical when observed; that's why this type is named "Elliptical". This galaxy has a complex central bulge containing countless surrounding stars, supermassive stars; specially several supermassive black holes or ordinarily only one. The complexly compact circular central core is spherically surrounded by shining stars, superluminous stars, supermassive stars with wonderfully separated star systems. These are authentically most massive gigantic galaxies of continuously constant Cosmos.

Elliptical galaxy is classified based on its spherically surrounding shape's stratifications and as its spherical shape changes from circularly spherical to elliptically spherical, its type changes comparably. Elliptical Galaxies are noted with latter "E" and to classify them the numbers "0 to 7" are used after the letter "E"; those numbers represent the degree of ellipticity; as the numbers change from 0 to 7, the shape of the elliptical galaxies change from perfectly circular sphere to elongated elliptical sphere. The common classification is described below.

**E0:** Correctly completely circular spherical shaped surrounding structure.
**E1:** Completely circular spherical shape structure but less than the previous type.
**E2:** Comparably circular spherical shape structure but less than the previous type.
**E3:** Comparably circular spherical shape structure but less than the previous type.
**E4:** Elongatedly elliptical spherical shape but less relative to the type noted next.
**E5:** Elongatedly elliptical spherical shape but less relative to the type noted next.
**E6:** Entirely elliptical spherical shape but less relative to the type noted next.
**E7:** Excellently entirely elliptical spherical shaped surrounding structure.

## Other Types of Galaxies

### Irregular Galaxy
Galaxies of randomly irregular shapes or peculiar shapes are Irregular galaxies; Large or Small Magallanes Clouds are also considered Irregular galaxies; these clouds are also notable Nebulas.

### Dwarf Galaxy
These are very small size galaxies relative to other normal galaxies.

### Ring Galaxy
These remarkably rare galaxies are ring-like galaxies; this round ring structure shiningly surrounds the central compact core of the Ring Galaxy.

# Formation; Evolution of Galaxies

Disk Galaxies or Spiral Galaxies form from supermassive superlarge gas clouds; then those spiral galaxies merge together to faultlessly form massive Elliptical Galaxies; this phenomenon is known as galactic evolution.

## Disk Galaxy Formation

The formation of Disk Galaxy which is also a Spiral Galaxy is still not totally understood but there are logical explanations. The most common explanation is that the disk galaxies formed from superlarge supermassive gigantic gas clouds. The gravitational collapse of a gigantic gas cloud gives it angular momentum; then the entire gas cloud starts spinning fast and also cause continuous cooling; after cooling down it emits energy to completely concentrate toward the center and because of conserved angular momentum more matters near the center spin faster; then that tremendous spinning gives it a disk-like shape. As it cools down more, the gaseous matters in different regions of the cloud disk, break apart and then separately cause complete gravitational collapses to themselves, to form shining stars; star systems.

## Merging of Galaxies; Evolution; Elliptical Galaxy Formation

Two or more disk shaped galaxies merge and form a massive Elliptical Galaxy. Galaxies bounded by gravity cannot escape evolution or galaxy merging; so they evolve by merging into a more massive elliptical galaxy. All gravitationally bounded galaxies will mergingly evolve to more massive extremely enormous Elliptical Galaxies.

# Star's Systematic States; Standard Stellar Evolution

In here we will discuss the evolution process of stars from formation to future fate.

# Star's Starting State

### Initial, Low Mass Protostar
An initial state low mass Protostar begins its formation from a significantly small separated segment of a large massive molecular cloud. The gravity of that small segment continuously concentrates most matters into its common center; compressingly combine those matters in the center to form a low mass Protostar.

### Brown Dwarf; Sub Stellar Objects
When a low mass Protostar fails to gain more mass to increase its total mass above 0.08 solar mass; it cannot gain enough temperature to start Hydrogen Fusion; so it never evolves beyond its initial stage and always remains as a Brown Dwarf or Sub Stellar Object. Brown Dwarfs usually fuse Deuterium instead of Hydrogen.

### High Mass Protostar
A low mass Protostar attracting more matters because of its gravity and abundance of surrounding gaseous matters, is turning time to time itself intensely into a relatively more massive Protostar or High Mass Protostar; this perfect Protostar will never remain as a Brown Dwarf but become a shining Star capable of fusing Hydrogen to emit extremely enormous electromagnetic emission; lively luminous light.

**Pre Main Sequence Star**

This is the previous stage of main sequence star; in this stage, the relatively high mass protostar attracted all of its necessary mass to start hydrogen fusion but still not started it because of necessary temperature. In this process; perfect protostars immensely increase their total temperature to totally turn themselves to Main Sequence stars, shining stars capable of fusing hydrogen.

# Main Sequence Star

Finally, the High Mass Protostar perfectly passed previous phase of stellar evolution to rightfully reach the triumphant temperature to start stable natural nuclear fusion for emitting extremely enormous electromagnetic emission; eminently enlightening energy; lovely luminous light. When the core temperature of newly formed stars reaches ten million Kelvins; they start to fuse Hydrogen into Deuterium and then to Helium; they continue this fusion process perfectly and brilliantly become Main Sequence stably shining stars. Then this constantly continuous process of fusing Hydrogen into Helium can completely reach Hydrostatic Equilibrium; in this star's state of equilibrium, energy emission of the star's core can completely maintain a high gas pressure; perfect pressure, that perfectly balances with working weight or its gravitational attraction toward center, to perfectly prevent gravitational collapse. After achieving Hydrostatic Equilibrium those stars evolve very fast and also become perfectly stable main sequence stars. Stars remain as stable Main Sequence Star for a very long period of time; the shining Sun is also a main sequence star; most of the stars of the Unlimited Universe are actually main sequence stars.

The time period of the main sequence star state depends mainly on star mass; Red Dwarfs with relatively low mass remain in this state for more than hundreds of billions of years; relatively mid-massed, mid-sized Yellow Dwarfs like the Sun will remain in this state for ten billion years; relatively more massive Red Giants remain in this state for only few millions of years.

"Most main sequence star's star systems can capably host habitable planets perfectly; seriously saying, so remarkably remember; hugely honorable humanity's habitable home "Eminent Earth" is immensely important; indeniably invaluable and also a profoundly perfect planet; outstandingly orbiting the tremendously marvelous main sequence stable star, significantly shining Sun; forcefully fusing Hydrogen Helium; eminently emitting enormous electromagnetic energy; encouragingly ensuring extremely essential honorable human's aliveness and also assuring all advantageous animal's aliveness and largely living lovely lives."

# Beyond Main Sequence State

## Least Massive Small Sized Stars

Evolution of small sized stars such as Red Dwarfs will be discussed in this section.

### Mass of 0.1 Solar Mass or Less

These least massive significantly small stars stably stay on the Main Sequence state; slowly increase their temperature and luminosity for more than hundreds of billions of years; then these stars slowly compressingly collapse into significantly small White Dwarfs without becoming Red Giants. This type of significantly small stars is not massive enough to generate a helium core with a surrounding shell that fuses hydrogen; so it will continue fusing Hydrogen until the entire star body becomes a body of only Helium; a star without a helium core cannot become a Red Giant Star; so it gravitationally compresses itself to directly become a White Dwarf.

### Mass More Than 0.1 to 0.8 Solar Mass

These significantly small stars can commonly reach to "Red Giant State" but cannot continue further toward the final phase of "Red Giant State", because their insufficiently massive helium cores cannot achieve enough temperature to fuse helium; so instead of reaching the final red giant phase they slowly decrease temperature, luminosity; gravitationally collapse themselves to become White Dwarfs.

# Less Massive Mid-Sized Stars

In this section, we will discuss the Stellar Evolution of mid sized Stars with mass from more than 0.8 to 8 solar masses. These stars significantly become Red Giants but cannot go Supernova.

**Subgiant State**
Mainly, main sequence stars enter this phase when all the hydrogen within their core is completely exhausted because of continuous fusion process; so these subgiant stars started fusing hydrogen within their Hydrogen Shell surrounding the Star Core. In this phase, the hydrogen shell produces more helium because of its hydrogen fusion; so the total mass of the helium core continuously increases; causing the star body to expand in size; decrees in temperature and luminosity relative to main sequence state; this process continues for multimillion to two billions of years depending on the initial mass of helium cores; then the helium cores degenerate for relatively less massive stars such as the Sun and for relatively more massive stars the outer layers continuously cool down to immensely increase layer's opacity; because of those constantly continuous changes those stars with their hydrogen shells increase in temperature, luminosity and size to expandingly enter extreme Red Giant State.

**Red Giant State**
In this state, stars hydrogen shells fuse more hydrogen and produce more helium; so the helium cores immensely increase in mass and also produce more luminosity; those more massive helium cores cause more rapid fusion of hydrogen within hydrogen shells continuously and as a result, those stars expand extremely to become greatly large; luminous Red Giants.

In this state, the core and outer layer of a star both expand relatively and become Convective or rapidly transfer heat energy because of bulk movement of gas particles; As that star grows more and more its shell's energy production increases rapidly and its outer convective layer expandingly goes deeper and deeper until it reaches deep enough to bring fusion products such as Helium from the convective core to the surface layer to increase the abundance of fusion elements in its surrounding star structures; star surface.

The extremely expanding helium core is no longer in thermal equilibrium; so it degenerates if that star is relatively less massive but for a more massive star its nondegenerate core can cause compression to reach the maximum mass limit of non-fusing helium core that can support surrounding star structures, layers.

## Final Phase of Red Giant State

The Helium Core is immensely increasing in temperature and also causing rapid fusion of hydrogen in the Hydrogen Shell; so the star's luminosity is increasing to brightest level to reach the tip of Red Giant Branch. A Star with degenerate core reaches the tip of Red Giant state when it increasingly achieves the sufficient temperature to start helium fusion in the "Helium Core"; that type of stars has a growth and luminosity limit and it reaches this limit in the tip of Red Giant State; this type of "Red Giant" stars shows significantly similar luminosity of two solar luminosities and temperature of three thousand Kelvins and also achieves an identical helium core of 0.5 solar mass. Any relatively more massive star with a nondegenerate Helium Core actually achieves enough temperature and also start fusing helium before reaching the tip of "Red Giant" state.

## Horizontal Stellar State

The Horizontal state starts after the complete completion of Red Giant State; then the degenerate helium cores of stars significantly start helium fusion in this stellar state of evolution. Stars with mass of two solar masses or less have degenerate Helium Cores; in this case, the starting of helium fusion within these cores immensely increase the temperature and fusion rate; then those degenerate cores completely become nondegenerate and also extremely expand by causing extremely explosive effect of Helium Flash, an after effect of Triple-Alpha process of fusing three Helium Nuclei into a heavier Carbon Nucleus. Helium Flash emits extreme energy, equal to $10^8$ times the solar luminosity but the plasma layers covering cores consume it completely. These stars are now in new equilibrium state; so they will stay in this "Horizontal State" for hundred millions of years.

Stars with mass above two solar masses have nondegenerate helium cores; these cores fuse helium smoothly slowly with no helium flash but these cores completely ensure enough temperature to fuse helium and also begin to fuse helium in its "Red Giant State" before becoming degenerate cores completely. These stars significantly cause a "Blue Loop" in its Horizontal State to evolve further; a blue loop is a stellar evolutional process that causes a significant star to become hotter from a relatively cold condition, before cooling down again. Relatively more massive stars cause more blue loops that can cause luminosity of thousands of solar luminosities. Stars simply stay in this Horizontal State while increasing in size and luminosity until they eventually exhaust every helium inside inner "Helium Cores".

## Asymptotic Giant Branch, State

Stars enter the Asymptotic State when they exhaust all the helium in their Helium Cores. After exhausting all the helium, the remaining is a hot core of carbon and oxygen; that oxygen-carbon core is surrounded by structural shells of helium and hydrogen. In this state of stellar evolution, star's fusion processes completely continues in its helium shell and hydrogen shell to rapidly produce extreme electromagnetic energy for shorter periods of time, depending in its main sequence state's mass; because of hydrogen fusion in the hydrogen shell all resulting helium falls toward the center; causes more rapid fusion of helium in the helium shell; it increases its electromagnetic energy emission extremely; effecting thermal pulses; these energy pulses occur near the end of the Asymptotic Giant State. This state decreases mass; stars lose fifty to seventy percent of mass in Asymptotic Giant Branch. This stellar state has two sub-states, Early Asymptotic Giant Branch and Thermally Pulsing Asymptotic Giant Branch. In Early State, the helium shell surrounding oxygen-carbon core mainly continues fusion process until it exhausts entirely; in this state star's spherical size expand to the diameter of approximately four hundred and thirty solar radiuses. The Thermally Pulsing State start when helium shell fusion ends and hydrogen shell fusion starts; after ten thousand to hundred thousand years of hydrogen fusion, the helium layer builds up again and explosively emit energy as helium flash; causing luminous flashing of thousand solar luminosities. In this phase, deep convective zone forms and it brings central core's carbons to top surface of the star; this is known as "Dredge Up" and some significant stars can cause multiple dredge ups that can create Carbon Stars, largely luminous red giant stars with atmosphere mostly containing carbons. This state ends when stars exhaust all its fusion fuels for further forceful fusion process progressions.

**Final Fate; Dark Dwarf; White Dwarf; Planetary Nebula;**
The final fate of a relatively less massive mid sized significant star is a relatively cold White Dwarf surrounded by planetary Nebula, the remnant of a star after Asymptotic Giant State's completion. After exhausting all fusion fuels these stars cannot cause carbon fusion because of insufficient mass; so they contract again and again and also continuously create Superwinds of ordinary stellar stuffs to produce planetary natural nebulas; then the remarkable remaining are highly hot compressed central cores; then those cores continuously cool; completely convert intensely into wonderful White Dwarfs, definite deathless dwarfs. White Dwarf's Electron Degeneracy Pressure prevent it from further gravitational collapse but a White Dwarf can constantly continue cooling; change itself into indefinite Dark Dwarf, Black Dwarf.

# More Massive; Supermassive Stars

In here, we will discuss the evolution of massive stars; we will consider stars with mass above eight solar masses as massive stars; these types of relatively large sized spherical stars subsequently start Supernova; explosively emit enormously enlightening extreme electromagnetic energy; result remarkable remnants.

**Supergiant Stars; Special Superlarge Stars; More Massive than Eight Solar Masses**

A main sequence star with mass above eight solar masses can fuse heavier elements to become a special Supergiant and all main sequence stars with mass between ten to forty solar masses become brightly big Red Supergiant and also fuse heavier elements. These stars cannot remain as Red Supergiant; so they all explosively effect extreme Supernovae.

## Superluminous Supermassive Stars; More Massive than Forty Solar Masses

This type of extremely massive and enormously luminous stars can cause constant rapid stellar winds; those stars also lose more mass rapidly because of their extreme radiation pressure; so stellar winds with dramatic decrement of more masses continuously cause strong stripping of surrounding stellar structures, star's envelopes or layers. Because of those significant situations, superluminous supermassive superbly shining stars with more mass than forty solar masses fail to become Red Giants but they generally go Hyper-Novae or Super-Luminous Supernovae.

"Superluminous supermassive; superbly shining stars can correctly cause cataclysmically supreme Superluminous Supernovae; explosively emit enormously enlightening extreme electromagnetic energy; extraordinarily enlighten entire endless entirety, expandingly unlimited Universe, continuously constant Cosmos."

## Supermassive Star's Structure; Progressive Processes of Fusion

Massive stars with more mass and tremendous temperature give them the capability to fuse heavier elements; these are progressive processes like Hydrogen fuses to Helium; then Helium to Carbon; then to Neon to Oxygen to Silicon to Iron and some other heavy elements; then they finally forcefully fuse Ferrum, Iron.

These fusion processes continuously create shell structures or layers of different elements within a massive star, such as an Iron Core with a surrounding shell-like structure of Silicon; then a surrounding shell-like structure of Oxygen; then a surrounding shell-like structure of Neon; then a surrounding shell-like structure of Carbon; then a surrounding shell-like structure of Helium; then finally a surrounding shell-like surface structure of Hydrogen.

# Star's Supreme Supernova; Superluminious Supernova

As described before, supermassive stars fuse heavier elements including Iron; forcefully fusing Ferrum or Iron immensely consume continuous electromagnetic energy; so the continuation of Iron fusing process requires more energy than it can create or the fusing, nuclei combining process creates less energy than the energy needed to break them in the previous process from their previous parent nuclei; then those stars dramatically decrees in energy and also fail to continue further fusion process; then their cores exceed the maximum mass limit of a stable White Dwarf which is 1.4 solar masses. If a star's core is more massive than 1.4 solar masses; its Electron Degeneracy Pressure cannot support star's working weight anymore, against aggressively attractive centrally compressing governing gravitation; so that star's core compressingly collapses completely; cause cataclysmic catastrophe; called supernova and it happens because of gravitational core collapse of a supermassive star.

When a White Dwarf or Star Core becomes more massive than 1.4 times the mass of the Sun; its Electron Degeneracy Pressure completely fail to prevent gravitational core collapse and as a result, it goes supernova; that circumstance can obviously occur when multiple White Dwarfs merge to a massive one. Stars with mass of forty solar masses or more can commonly cause Superluminous Supernovae or Hypernovae; these extreme explosions are more than ten times luminous than standard supernovae.

## Classification of Supernova

Supernovae are classified based on their light curve or absorption line in their spectra; light curve is the observed luminosity relative to time and spectra is the electromagnetic spectral lines of any object and absorption lines or dark lines within the spectral lines occur because that object's composing elements absorb some wavelengths of particular electromagnetic energy, color lines of light.

## Type I

This type of Supernova shows no Hydrogen absorption line; it's subdivided into:

**Type Ia:** This type shows strong Silicon significantly ionized, absorption line; it indicates the profound presence of ionized silicon.

**Type Ib:** This type shows weak or no silicon absorption line but show Helium's line; it indicates the profound presence of Helium.

**Type Ic:** This type shows weak or no silicon or helium absorption line.

## Type II

This type of Supernova shows strong Hydrogen absorption line that indicates the profound presence of Hydrogen; it's subdivided based on the light curve into:

**II-P:** Show no narrow line and reaches to plateau in its Light Curve.

**II-L:** Show no narrow line and show linear decrement in its Light Curve.

**IIn:** Show some narrow line in its Light curve.

**IIb:** This type actually shows features of Type "II" and "Ib"; it shows Hydrogen absorption line initially but then after some weeks or months it shows helium absorption line like Ib.

## Type-Pec

Any supernova failing to classify itself into common classification described above will be classified as Type-Pec; this "Pec" means peculiar and all supernovae showing peculiar properties are categorized as Type-Pec or as peculiar type II.

## Superluminous Supernova or Hypernova Classification

Superluminous Supernovae or Hypernovae occur when super sized supermassive stars with mass more than or equal to forty solar masses explode; these explosions are ten times or more luminous than standard supernovae; some superluminous supernovae are hundred times luminous than significantly standard supernovae. Hypernovae have different classification; they are described below.

**SLSN I:** This type of Super-Luminous Super-Novae shows weak or no Hydrogen absorption line; it indicates poor presence of Hydrogen. This type's light curve is comparable to Supernova type Ic; so this type also classed as SLSN-Ic.

**SLSN II:** This type of Super-Luminous Super-Novae shows strong Hydrogen absorption line; it indicates profound presence of Hydrogen.

**SLSN III:** This type of Super-Luminous Super-Novae shows no Hydrogen absorption line and also show peculiar luminosity and powered by Nickel.

# Superbly Supreme Supernova's Remarkably Resulting Remnants

In this section, we will discuss the stellar remnant of stars after going Supernovae.

### Neutron Star

These are the densest stars of the Unlimited Universe and we discussed these objects before. White Dwarfs or Star Cores exceeding the mass of 1.4 solar masses but not 2.2 solar masses, remarkably remain as Neutron Stars after Supernova explosions. When a star is more massive than 1.4 solar masses it collapses more by causing electron capture, this is a particular process of a proton rich nucleus of an electrically neutral atom, absorbing an atomically inner Electron, that orbits in inner orbital position close to the nucleus.

The continuous Electron Capture processes reduce the Electron Degeneracy Pressure of a Star Core; so it continuously collapses forcefully further because of its gravity, to become an extremely dense star known as Neutron Star. A Neutron Star spins very fast almost over six hundred revolutions per minute.

## Black Hole
The most mysterious object of the universe is a Black Hole; White Dwarfs or Star Cores with mass above 2.2 solar masses cannot cause enough Electron Degeneracy Pressure; prevent further forceful greatly gravitational core collapse and always become black holes; hence commonly cause cataclysmically supreme supernovae; sometimes special cases can cause notably no supernova, surprisingly. The tremendously mysterious; marvelously terrific thing of any authentic black hole is its inside supreme singularity surprisingly surrounded by escapeless event horizon, here time totally stops; superbly superfast leaving light fully fail to travel; longingly leave.

## Resulting Remnant; Core Collapse Classification

### Electron Capture Caused by Oxygen; Neon; Magnesium Cores
Stars with mass between eight to ten solar masses are in this category; their typically collapsing cores cause faint Type-IIp Supernova to become Neutron Stars.

### Pair Instability
This is a special case of supermassive stars with low metallicity and mass between 140 to 250 solar masses; these star cores pair production; producing free positrons and electrons, effects pressure; prevents gravitational core collapse but when their pair production pressure plummet, these tearing cores cause Type II-P Supernova or Hypernova with gamma burst but no resulting remnant.

## IRON CORE COLLAPSE

### Star Mass between 10-25 Solar Masses
Stars with mass between ten to twenty-five solar masses with collapsing cores cause faint Type II-P Supernova to result Natron Stars.

### Star Mass between 25-40 Solar Masses
Star cores with low metallicity cause normal Type II-P Supernova to initially become Neutron Stars with surrounding star stuffs; then those Neutron Stars gravitationally consume surrounding star stuffs to become Black Holes.

Star cores with very high metallicity cause Type IIb or Type II-L supernovae to become Neutron Stars.

### Star Mass between 40-90 Solar Masses
Star Cores with low metallicity; they cannot cause Supernova; so they directly become Black Holes.

### Star Mass Equal or Above Forty Solar Masses
Star cores with near solar metallicity cause Faint Type Ib, Ic or Hypernova with Gamma Ray Brust; then they initially become Neutron Stars with surrounding star stuffs; then those Neutron Stars gravitationally consume surrounding star stuffs to become Black Holes.

Star cores with very high metallicity cause Type Ib or Type Ic supernovae to become Neutron Stars.

### Star Mass Equal or Above Ninety Solar Masses
Star Cores with low metallicity; they cannot cause Supernova but cause gamma ray burst; then they directly become Black Holes.

## Photodisintegration Process
This is a special case for supermassive stars with mass more or equal to 250 solar masses. Photodisintegration process works within condensed Cores of these stars; this is a nuclear process of atomic nuclei absorbing high energy gamma rays to become energetically excited; then they immediately decay by emitting subatomic particles. This photodisintegration process profoundly absorbs energy to remarkably reduce temperature and pressure of those cores that causes gravitational core collapse; cause no Supernova or possibly cause Hypernova but always become massive Black Holes.

# Supreme SpaceTime Singularity; Incalculably Immeasurable; Inestimably Indeterminable; Illimitable Informational Infinity

**Mathematical Singularity**
We humans, use mathematics as a tool to creatively construct theories of physics; mathematics merged wonderfully with physics, perfectly predicts physical phenomena; correctly calculate cosmological circumstances; countless cosmic entities; elegantly express endless entirety, unlimited Universe. If meaningful mathematics cannot calculatingly define a profoundly particular point, then that point is a mysterious mathematical singularity; All mathematical calculations completely become incalculable; indeterminably infinite in that particular point of singularity.

**Relativity's Gravitational Singularity; Spacetime Singularity**
Space-Time Singularity or Gravitational Singularity is a one-dimensional point of space-time of a complex cosmic object, where the gravity is infinite because of the infinite density of that object.

**SpaceTime Singularity's Starting Source**
Any Black Hole contains a single spacetime singularity in its exactly correct central coordinate of spacetime; when a star core or white dwarf is more massive than 2.2 solar masses it gravitationally collapses completely to cause complex spacetime singularity; then that spacetime singularity causes an Event Horizon because of its extreme gravitation; a singularity with an event horizon is a Black Hole.

## Physical Properties

Gravity is spacetime curvature; at the exact point of a singularity or in the center of a Black Hole, where its singularity exists, the curvature of spacetime is infinite or the maximum depth of spacetime curvature is infinite; at that point, the spacetime curvature extends indefinitely into infinity. Theory of gravity breaks down in the exact point of a singularity because of its infinite gravity caused by infinite density. Event Horizon is not an object but a gravitational field boundary, where the escape velocity is equal to light speed; so no light particle, Photon can ever escape from that boundary because of gravitational acceleration toward the center; light fails to travel outward from the Event Horizon, even with enormous velocity; so the inside of every Event Horizon is unobservable; the area enclosed by Event Horizon is dominating darkness.

## Event Horizon; Luminous Light

Event Horizon of a singularity is relative to light; the event horizon is the area of complete darkness because no light can ever return to the observer from the event horizon and absence of light or no returning of light cause complete darkness. If no light or light source is present around a singularity, then its event horizon is completely unobservable. the event horizon is the effect of light relative to light, caused by a singularity; if no light is present then the event horizon cannot exist; even everything fails to exist without originally observable effects, evidences.

In circularly spherical surface of Event Horizon, the gravity is strong enough to prevent light from escaping; as an imaginary point keeps moving toward the singularity by crossing the event horizon; the gravity keeps increasing, spacetime curvature increasingly ingoing into infinity relative to surrounding spacetime; when that point reaches the exact location of the singularity; it actually exists in a location where the gravity is infinite; its spacetime curvature extended endlessly into indefinite infinity. The Supreme Singularity is the most gravitationally powerful object because of its infinite gravity.

**Cosmic Singularity;**
**Supremely Surrounding Spacetime's Significant Starting**
This is beyond Big Bang; the supremely superior Singularity of the completely compressed Cosmos caused the Big Bang that gave birth to the entire endless entirety, unlimited Universe. Before Big Bang, time and space, Spacetime and every energy and all matter's mass exceptionally existed within a complexly compressed tiny thing of infinite density; that tiny thing is the Cosmic Singularity.
The four fundamental forces also existed within that singularity as one force that only caused infinite gravity. The three forces except gravity; the Electromagnetic, Weak Nuclear, Strong Nuclear forces have elementary particles; so they can combine to become one in extreme energy environment of high temperature. Those three forces combined to become one; then that combined force capably combined completely with gravity to become only one that only caused gravitation; that one unified force and all surrounding spacetime and all mass, matters, energies entirely existed as a completely compressed indefinitely infinite dense point of Singularity; this is the explanation of Cosmic Singularity by going backward in time toward the Big Bang; beyond Big Bang; before Big Bang's bright beginning; before Big Bang's brilliantly breathtaking birth.

"The Time Traveler, tremendously traveling time to time; traveling the total time backwardly; beyond the time's brilliant birth; before Big Bang's beginning; before surrounding spacetime; before four fundamental forces; before mass, matters, energies, every entity, every emerging existence; entirely established entirety exceptionally existed wonderfully within complexly compressed Cosmic Singularity; then that supremely superior singularity cataclysmically caused Big Bang; birthingly brought entire endless entirety, unlimited Universe into its emerging existence."

# Complete Cosmic Creation; COSMOS

In here, we will discuss the formation of the Universe from Big Bang. Some sub sections are in present form because the series of phenomena expressed in those sections happened in a very short period of time.

**Planck Length; Planck Time:** The time light needs to travel one Planck Length, **$1.61 \times 10^{-35}$** meter, is one Planck Time, **$5.39 \times 10^{-44}$** second; In Planck Scale, the Quantum Effects become dominant realities; the Planck Length and Planck Time are the smallest universal units of Length and Time. Now, we will impart important information to flawlessly form the Unlimited Universe.

## Big Bang

First, there was a cosmic singularity, a one-dimensional point that extraordinarily evolved to the tremendous temperature and infinite density state; that cosmic singularity was beyond any spacetime singularity of any black hole in the unlimited universe; we can calculably understand or observe. All the surrounding spacetime, mass, matters and energies existed wonderfully within that terrific cosmic singularity; then it goes Bang, an extreme explosion that gave birth to space-time, mass, matters and energies of the unlimited universe, we observe now. After a very short period of Big Bang, the early universe, early entirety expanded extremely, more than the light speed to take its initially small sized Spacetime Structure; this is what happened about 13.82 billion years ago, before the existence of Time and Space of the Universe, we live.

## Initial Inflation

After the tremendously extreme explosion of Big Bang, the early Universe or early entirety's existence is extremely expanding and also continuously cooling relative to its initial state, for approximately $10^{-43}$ second or 1.9 Planck Time and after that time of extreme expansion, the Cosmic Inflation will begin to cause continuous effects. Within this profound period of cosmic inflation, the early entirety's existence is expanding exponentially and because of the uncertainty principle of quantum mechanics, the Quantum Fluctuation or vacuum state fluctuation, empty entirety's endless energy essentially emerging everywhere in every point of the early Universe. Energy can convert itself into mass; so this quantum fluctuation or energies of empty space can capably amplify itself into the seeds of large structures of the observable Universe; we are observing in this time.

This is how our own Universe formed from energies of empty space or notably nothingness. The quantum fluctuation is the continues change in the amount of energy in any point of spacetime and this was described before. Now, we will go to the next phase after the Inflation Period.

**Particles Phase**
After only $10^{-31}$ second, the inflation period ended; the total temperature of the early Universe is now hot enough to form the Quark-Gluon plasma, a highly hot compactly compound perfect plasma of elementary particles, a particular plasma of elementary Quarks and Gluons; Quarks are the elementary particles for matters and the Gluons are the elementary particles of Strong Nuclear force. The early Universe is now in perfect pressure; total temperature for the Quark-Gluon Plasma to form; so it's forming faultlessly.
Within this particular phase of the early Universe, all the other elementary particles are also forming faultlessly; so now the early universe has the elementary particles of energy or Photons and also all the elementary particles to form mass, matters but the total temperature and density of the early Universe are still extremely high to form any type of mass. So, Let the Plank Times pass perfectly.

**Baryon's Birth**
As the Plank Times are passing, the early Universe is cooling continuously and also dramatically decreasing its immense density; so the energy of the elementary particles is also decreasing dramatically; then the systematic symmetries are breaking and the four  fundamental forces of perfectly proven physics and all the particular properties of the elementary particles are perfectly progressing to take their present particular form or in other words, the theories of the presently proven physics of the observable universe are becoming universal facts.

After only $10^{-11}$ second of continuous cooling and energy decreasing, particle energies of the early universe are reduced to the lower level like the particle accelerator's achievable energy level; then after only 0.000001 second, the early universe is now in the perfect energy state to form Baryons; so together, the quarks and gluons of the early universe are flawlessly forming Baryons. Baryons are composite subatomic particles such as Protons and Neutrons; nearly all the matters are made of baryons; so these matters are called Baryonic Matter.

**Continuous Cooling**
After only 1 Second, the temperature of the early universe cooled enough to give survivable stability to the positive particles and Baryons. Within this period the photons are becoming energy dominators of the early universe; they are dominating the energy density of the universe. After some minutes the total temperature of the early universe is approximately one billion Kelvins and its density is almost as the density of normal air and then the baryonic particle Neutrons and Protons are capably combining to form Deuterium and Helium nuclei. The Deuteriums are Hydrogen's heavy Isotopes. Time to Time, the total temperature of the early emerging Universe is decreasing dramatically and so, the Rest Mass energy of matter is gravitationally dominating the early universe, which was previously dominated by Photons; these progressive processes will take approximately 378,000 years to perfectly complete.

**Amazing Afterglow; Cosmic Microwave Background Radiation; CMBR**

After only 378,000 years in the early Universe, the existing elementary Electrons and new natural Nuclei are combiningly creating historic Hydrogen Atoms and emitting enormous electromagnetic emissions everywhere enlighteningly; this early energy emission is the Cosmic Microwave Background Radiation or CMBR; these remarkable radiations are the afterglows or the triumphant foundational framework from flawless formations of entire endless entirety, ultimately unlimited Universe. After amazingly apparent afterglow of the early universe, the definite dark age of the unlimited universe will begin; as the times are passing the temperature of the early universe is dropping dramatically and all relatively denser regions of matters are gravitationally attracting their surrounding matters to become more massive; so their densities are increasing more; then those more dense regions are attracting more matters to become more and more massive; these profoundly progressive processes are authentically progressing perfectly to flawlessly form marvelously massive magnificent objects of constantly continuous colossal Cosmos, like light emitting entities or superbly shining stars.

These progressive processes are historically happening wonderfully within almost all remarkable regions of supremely surrounding spacetime of early emerging endless Entirety; these processes will take very vast period of times to faultlessly form first fantastically shining stars; so the Dark Age of the early Universe will constantly continue until the brilliant birth of the first fabulously significant shining star.

## Shining Star's Significant Starting

Approximately two hundred and fifty million years after the Big Bang, the first shining Star is brightly born brilliantly; this is the superbly shining star that ended dramatically dominating darkness of the dominating Dark Universe wonderfully, with its immensely luminous lovely light by emitting electromagnetic emission, early enlightening energy. Approximately, four hundred million years after the Big Bang, the first fabulously gorgeous Galaxy forms fantastically; like this, other stars and galaxies are forming too, Time to Time. The Triumphant Time, perfectly passing continuously; countless cosmic creation's creational processes perfectly progressing enormously everywhere; early emerging Entirety's everywithin, every region of early emerging Universe. The tremendously governing gravities of remarkable regions with dominating density of massive matters are attracting other surrounding matters to become more massive and also increasing in density; then together those massive masses are forming early Gas Clouds, Stars, Planets, Moons, Black Holes and Galaxies in every region of the early Universe to generally give it its current colossal configuration; flawless form. There are greatly gigantic supreme structures in the unlimited universe than the gigantic galaxies; we will definitely describe them too.

# Colossal Cosmic Configuration; Significantly Supreme Structures

The Mass of the observable Universe is approximately $10^{53}$ kg but this is just the mass of ordinary macrocosmic matters; the planets, stars, galaxies, interstellar medium, intergalactic medium and all ordinary observable objects are made of ordinary matters. In the universe there are things beyond ordinary matters and energies; they are Dark Matter and Dark energy; these things are not directly observable but their effects are obviously observable. The Universe contains seventy percent Dark Energy; twenty-five percent Dark Matter; five percent ordinary matters made of ordinary Atoms. The Dark Energy causes the accelerating expansion of the entire universe; the mass of Dark Matter causes most of the gravitation of the universe to bound the galaxies; galaxy groups; the observable universe together.

**Dark Matter**
The dark matter is not ordinary matter and it is beyond the standard model; this matter is invisible; interaction less; causes no interactions with light, ordinary matters but constantly cause gravitational effects and affects everything everywhere entirely.

**Galactic Gravitation**
Within a galaxy, all the stars orbit around with similar velocities because of galactic gravitation; the orbital speed of stars does not decrease relative to distance from galactic center like the star systems; the stars near the galactic center and the stars near the galactic edge orbit with similar speed or in common case the stars in the edge orbit with slightly greater velocity relative to the orbiting stars near the galactic center.

This is also explained in the "galactic rotation curve" of galaxies; it's a graph of star's orbital velocities relative to their distances from galactic center; in this graph, the star's velocity curve continuously increases relative to the star distance from center. These phenomena happen because of the extra mass of the dark matter within galaxies; the gravity caused by the mass of Dark Matter causes relatively similar orbital speed of stars within galaxies or causes the observed galactic rotation curve.

**Galaxy Group's Gravitation:** The observed distances between galaxies within gravitationally bounded groups are very vast but still they manage to gravitationally attract each other to keep themselves together in groups; the observed distances between bounded galactic groups are also very vast but they are attracted to each other and bounded by gravity together, because of the gravity field caused by the total mass of dominating Dark Matter.

**Dark Energy**
We already described that the Dark Energy is the energy of empty space existing in every point of empty space to constantly cause the accelerating expansion of the universe. This accelerating expansion is perfectly proven by observing the luminous supernovae and the motions of further galaxies in the Universe. The accelerating expansion rate of the observable universe is seventy-three Kilometers per Second per Megaparsec, Hubble Constant; it means that the universe is acceleratingly expanding relative to distance; for every megaparsec distance, the accelerating expansion rate of space is seventy-three kilometers per second; this increases relative to per megaparsec distance from the Earth.

## Blue Shift; Red Shift of Galaxies
Different colors of light have different wavelengths; the blue light has shorter wavelength than red light. In the universe, when a galaxy is in a motion toward the Earth; its luminous light becomes blue because of the compression of its wavelength or in other words, its motion toward us causes a shortening of light's wavelength because of the wave properties of light; like this the galaxies moving away from us appear emitting red light because of its opposite directional motion, extending the wavelength of its light. The more further we observe in the universe, the galaxies are observed more red, redshifted relative to their distances from us; that confirms the endless expansion of unlimited universe.

## Some Significantly Outstanding Objects of Observable Universe
Our Sun is made of 70 percent Hydrogen and 28 percent Helium and 1.5 percent Oxygen, Nitrogen, Carbon and 0.5 percent Neon, Iron, Silicon, Magnesium and Sulfur. The Sun is very big relative to the Earth and approximately 1.3 million Earth can fit inside the Sun but in the observable universe, there are larger stars than the Sun. The NML Cygni, a hypergiant star; its radius is 2,800 solar radiuses; its mass is 50 solar masses; it is as luminous as 300,000 Suns. The UY Scuti, a hypergiant star; its radius is 1,800 solar radiuses; its mass is 10 solar masses; its luminosity is 350,000 solar luminosities. The VV Cephei, a hypergiant star; its radius is 1,500 solar radiuses; its mass is 18 solar masses; its luminosity is 200,000 solar luminosities. VY Canis Majoris, a hypergiant star; its radius is 1,500 solar radiuses; its mass is 23 solar masses; its luminosity is 300,000 solar luminosities.

The largest galaxy in the observable universe is the "IC 1101"; this galaxy is approximately four million light years in diameter; this galaxy contains almost one hundred trillion stars. The most massive black hole in the observable universe is the hypermassive black hole in the galactic center of IC 1101 galaxy; this hyper-massive black hole is as massive as one hundred billion solar masses.

## Intergalactic Dark Voids

Gas Nebulas, Stars, Planets, Moons, Asteroids and Black Holes together form the Galaxies but galaxies are not the largest structure of the observable Universe. There are dark voids of space in the universe, which are greatly larger than gigantic galaxies; if we consider them things or single space things then they are the largest things in the observable universe. Space void is not an object but the vastness of unoccupied space within a region of spacetime; there are large regions of space where no galaxy exists and those dark regions of space are dark voids or space voids.

## Surrounding Structures of Observable Universe

The observable universe is simply a complete combined collection of greatly gigantic groups of gigantic glowing galaxies; these galaxy groups together formed the large-scale structures of the observable universe; then those super structures together formed the supreme structure of the entire observable universe.

## The Local Group; small galaxy group

Galaxies together combiningly form group of galaxies because of the gravitational attraction between them. Galaxy groups are larger structures than galaxies. The Milky Way our home galaxy and the Andromeda galaxy and the Triangulum, together exist within a galaxy group called the Local Group.

The Local Group contains more than fifty-five galaxies and also astronomical objects such as gas nebulas and magnetic clouds. These compound cosmic collections such as galaxy groups are particular parts of the large scale structure of the observable Universe. The Local Group has a diameter of approximately 3.1 *Mpc*, megaparsecs, 10.11 million light years.

### Galaxy Cluster; Supercluster

Galaxy groups together form a gigantic galaxy Cluster; these clusters are even larger than galaxy groups like Local group. The Virgo Cluster is a cluster near the local group; the Virgo cluster and Local Group along with some other galaxy groups are members of a Supercluster called the Virgo Supercluster and approximately more than hundred galaxy clusters and galaxy groups exist within the Virgo Supercluster; this supercluster has a diameter of approximately thirty-three *MPC*, megaparsecs, 107.6 million light years. The Virgo Supercluster is also a part of a greatly larger structure of galactic clusters called the Laniakea Supercluster or Hypercluster; this greatly gigantic galaxy cluster has a diameter of approximately 160 *MPC*, megaparsecs, 522 million light years. Approximately ten million superclusters exist in the observable Universe.

### Greatly Gigantic, Galaxy Filament

There are structures larger than superclusters of galaxies; these giant structures are called the Galaxy Filament; these filaments are gravitationally bounded groups of separate supercluster of galaxies; these are called "Filament" because they will look like glowing filaments from a particular position of greatly vast distance.

The Laniakea supercluster is just a part of a galaxy filament named Pisces-Cetus Supercluster Complex, galactic filament; This galaxy filament is approximately one billion light years in length. The largest galaxy filament in the universe is Hercules-Corona Borealis Great Wall; its long length is approximately three Giga-Parsecs or ten billion light years.

## Complexly Colossal Collection, Cosmic Web

This galaxy filament "Pisces-Cetus Supercluster Complex" is adjacently adjoined alongside another gigantic galaxy filament called "Perseus-Pegasus Filament"; like this, some filaments form small sections; then many more filaments form supersized sections of a supreme sized structure, web-like framework. All those filaments are continuously connected to each other randomly, within a complex collection of filaments, by their own gravitational field; then together they formed a greatly gigantic; gorgeously glowing wonderful Web of Galaxy Filaments; this complexly connected glowing gigantic web of galaxy filaments is called Cosmic Web. The Cosmic Web is the largest structure of the universe; infact this Cosmic Web is the present form of the entire universe; if we can travel outside the universe to a greatly vast distance, then the entire structure of the universe will look like an arbitrarily adjoined and assorted giant glowing web; this web is the Cosmic Web. From an infinitely dense and significantly small particular point or Cosmic Singularity, the Universe expanded into a complete colossal Cosmic Web. The honorable humans live in a tiny habitable planet, which is actually smaller than tremendously tiny things relative to the greatly gigantic complexly connected colossal CosmicWeb; this is how significantly small the total humanity's habitable home is; this heavenly home is especially eminent "Earth".

"Constantly continuous colossal Cosmicweb; containing completely connected coruscating cosmic conglomerations; complexly consisting cosmically compound colossal clusters; containing complexly combined compound constellations; containing countless candescently colorful celestial cores, colorlessly caliginous cataclysmic canyons, countless celestially curved creations, cosmic creations; celestially curved creation continuously circling candescently colorful celestial core; capably containing cosmic consciousness, creaturely clever characters; cosmically coexisting cleverly contemplating constantly continuous ComicWeb, continuously constant Cosmos."

# Implement Infinite Imagination; Create Complete Cosmos

This is your infinite imagination; everything in here is only yours; only you can create; you can cause; you can complete cosmos by implementing infinite imagination. Every eye closed completely; this is your dark canvas; darkness is infinite; darkness is truth; darkness is freedom; in your sleep you exist in dramatic darkness; no matter how many; much love you have, you are lonely in sleep inside infinite darkness but you always authentically have youthful you; your heart; heavenly heartbeats; spiritually supreme soul; sufficiently surviving self. In the center of the imaginary dimension system or empty dark space of your imagination, create a cosmic singularity supremely. Even in your dark imagination, you cannot observe the cosmic singularity, because it's dark but you know it exists there, because you caused its creation. You created a cosmic singularity that can create complete cosmos; this is totally yours so only you can cause continuous cosmic creations by brilliantly beginning bright Big Bang; by breaking symmetries supremely; superiorly supreme is you inside indefinitely independent; illimitably infinite imagination of your marvelously magnificent mind. So, supremely cause cataclysmic Big Bang, an extreme explosion; then the early entirety extremely expanding exponentially; now give it its initial shape of spacetime and also cause cosmic inflation within that shape. The endless energy of empty spacetime is fluctuating within that initial shape; so amplify those energies to convert them into the seeds of marvelously massive supersized structures of unlimited universe.

This truly emerging early entirety of infinite imagination is continuously cooling and dramatically decreasing density and also decreasing the tremendous temperature while expanding exponentially. Now create the quark-gluon plasma in this early universe and also completely create all elementary photons and all essential elementary particles perfectly. Now dramatically decrees the total temperature to lower the total energy state; then the entirety's energy state is perfect to form Baryonic particles; so create baryonic particles by combining quarks and gluons to form baryonic matters. Now, decrees temperature more to stably create Hydrogen and Helium atoms and then continuously create countless Hydrogens and Heliums by combining baryonic particles, Protons and Neutrons and also cause creation of luminous light in that matter making progressive process; then cause those wonderful wave-particle Photons to travel toward the future with light speed; this is CMBR, an astonishing afterglow of the unlimited universe of infinite imagination.

Then, continuously combine more mass, more matters in all remarkable regions of surrounding spacetime of unlimited universe to create countless cosmic creations creatively; then create your first shining Star by compressingly combining more matters in a particular place; then creatively create countless gigantic galaxies and all celestially compulsory cosmic components.

Time to Time, the accelerating expansion is endlessly expanding this universe; the galactic gravitation is fascinatingly forming gigantic galaxy groups; then those galaxy groups are forming larger groups or galaxy clusters; then those colossal clusters are forming supermassive supersized superclusters; then those superclusters are faultlessly forming gigantic galaxy filaments and then those flawless filaments connectingly causing complete Cosmicweb.

It's your tremendously triumphant time to outstandingly observe by endlessly exploring your unlimited universe. Now, hold your universe with your imaginary highly huge hands to totally observe it; so you are observing the glowing Cosmic Web made of fiber-like glowing filaments; every point of light or pixel in those filaments is a glowing galaxy. Move toward the Milky Way with infinite speed and you are inside Milky Way instantly; it's a gorgeously glowing gigantic galaxy; then think the solar system's spot and you are instantly there; It's time for you to visit your own home planet Earth; so move toward it with infinite speed; you are instantly there; it's your heavenly home planet, perfectly eminent Earth; a beautifully blue perfect planet with wonderfully gorgeous greenness; the only obvious heavenly home of honorable humanity.

With the power of infinite imagination, you created the unlimited universe from a Cosmic Singularity and also endlessly expanded it to faultlessly form the CosmicWeb; then you observed that Cosmicweb. After completely creating; outstandingly observing your unlimited universe, you returned to your perfect planet; heavenly habitable home; excellently eminent Earth, honorable Humanity's obviously only one heavenly habitable home.

# Imaginary Information;
# Probabilistic Possibilities;
# Heavy Hypothesis

This section is a bonus section and we will mainly discuss imaginary information or hypothesis in here; we will discuss them to make you realize the power of infinite imagination but always keep in mind that imaginary information or hypothetical theories are just imagination not facts without observable evidences, scientific evidences; so you cannot accept imaginary information as universal facts without authentic serious scientific evidences. Because this book is mainly a scientific book, this hypothetical section will be simply small; if we write a book based mainly on imagination's infinite powers in the future; that book will definitely describe these imaginary ideas broadly and also beyond them.

## Multiverse; Many Many Unlimited Universes
This is a theory beyond human observation; this theory also has many subdivisions or possibilities. The multiverse is a complexly combined collection of unlimited universes like our own universe; our universe is an ordinary member of the multiverse. Different type of multiverses with different type of universes are described below.

## Multiverse of Infinite Universes: An expanding multiverse containing infinite universes similar to us with same physical laws and constants. Those universes are enclosed by separate Hubble sphere or universal bubble, a sphere of the observable universe where the observations come to an end. Universal Bubbles within the multiverse have so great distances between them, that even light cannot pass; so we cannot observe them.

**Multiverse of different Infinite Universes:** This multiverse is same like the first one but the universes within it have different physical laws and different constants.

**Multiverse; Many Worlds:** This is a multiverse predicted from quantum mechanics or a multiverse of different possibilities or all possibilities; in this multiverse, every possible result of any action has a separate universe; so this multiverse contains infinite similar universes of infinite possibilities.

**Quilted Multiverse:** A multiverse of infinite spacetime; all possible result of all actions are occurring here infinite number of times; infinitely.
**Inflationary Multiverse:** Within this multiverse, inflation occurring in every remarkable region infinitely and giving birth to infinite universes.

**Quantum Multiverse:** A multiverse of infinite universes; a new universe is created constantly for every possible result of every emerging event.

**Holographic Multiverse:** A multiverse containing holographic universes; those universes and all things within them are simply holograph from its initial source or consider it the Big Bang.

**Simulated Universe:** A multiverse containing simulated universes; consider that multiverse as an infinitely powerful computer and all the universes within it are just computer simulations or programs; objects within a program are authentically real to each other.

**Ultimate Universes; Multiverse:** A multiverse containing all mathematically possible universes of same and also different physical laws and constants.

**Multi-dimensional Multiverse:** A multiverse of infinite dimensions and universes within it have different number of dimensions, such as some universes have four dimensions and some have ten or eleven or more.

**Black Hole Multiverse:** A multiverse of infinite universes enclosed within infinite black holes and all those universes are existing within different separated black holes.

**Possibilities of an Existing Universe**
A universe has many different possibilities or final futuristic fates.

**Endless Expansion:** Expanding existing universe can constantly expand endlessly.

**Continuous Contraction:** Universe can stop expanding and then cause continuous contraction to become a singularity again and then it will go bang again; this is happening for infinite time.

**Terminal Temperature:** Endlessly expanding universe can cause all separated structures of the universe to travel to vast distances from each other; then they cool down and finally face dark death. A universe can also heat up completely to finally face its dramatic death.

## More Meaningful Multiverse

This is a theory of a multiverse of infinite expanding universes and it also explains the birth process of the universe according to observed model and proven physics. In this multiverse, within an endlessly expanding universe similar to us, all the different gravitationally attracted and attached galaxy groups cannot ever escape from each other and so those galaxy groups gravitationally merge to different separated spacetime singularities. While the galaxies within different groups are moving toward each other to form singularities, the different groups are also moving away from each other because of the accelerating endless expansion. When each of the groups becomes a complete singularity, the distances between separated singularities become extremely enormous. Now those singularities will remain as singularities for a very long definite period of time and within this time they will consume spacetime surrounding them and also its virtual particles and endless energies; then a particular moment comes when these singularities become so massive and consume spacetime so rapidly that the surrounding spacetime ends or fail to generate more virtual particles and endless energies; this is the moment when those cosmic singularities cause Big Bangs because of negative-positive mass explosions, energies and give birth to new universes of only positive mass. In this multiverse model any existing universe creates more universes and this progressive process continuously cause infinite universes.

## Singularity's Structure

A singularity can have many possible structures and they are discussed below.

**Point of Infinite Energy:** A singularity can be a single 1-dimensional point of infinite energy; energies do not interact with each other and infinite energy can stably stay in one point and cause extreme gravity.

**Orbit of Infinite Energy:** Like lights unstably orbiting outside a black hole, they can also orbit inside event horizon in a tiny point of space but to calculate that orbit the gravitational constant for mass and photon need to be obtained.

**Point of Empty Space:** The energy of empty space gave birth to the universe; the entire empty space can exist within a single point; so a singularity can be a point of empty space containing enormous endless energies.

**Point of Infinite Information:** A singularity can be a point of infinite information of spacetime, gravitation, waves, mass, energy, inflation, possibilities of events.

## Formation of Disk Galaxy

Disk Galaxies formed from predominant black holes; those black holes still remaining in the center of those galaxies. When predominant massive black holes formed in the early universe they attracted more mass, massive molecular clouds; then they stably orbited them and formed disk galaxies by accretion process. The region near those predominant black holes was denser and so they firstly formed massive stars to form a relatively small galaxy; then that small galaxy attracted more masses around them and also formed more stars; these progressive processes continuously occurred and also gave birth to many stars to form gigantic disk galaxies.

## Galactic Gravitation

Dark Matter is still undetected, unproven; it does not interact with ordinary matters but cause gravitation; the massless light's particle photons interact with ordinary matters but not with dark matter, even when it has massive mass. Dark matter causes gravity; so it must also be attracted by gravity but ordinary objects with gravity such as massive stars, planets are not attracting dark matter to consume it within them; to totally become unstable or possibly black holes. The hypothetical dark matter is unproven but it is the best explanation of different mysterious cosmic circumstances. The observed velocities of the stars within galaxies can also be calculated using Theory of Galactic Gravitation, obtained from Newtonian Gravitation for multiple bodies; this theory assumes an entire disk galaxy as an active accretion disk where anything in any point within the disk is gravitationally attracted toward the center or like the area of spacetime curve where every point is gravitationally concentrated toward the center. This theory calculates the gravitational acceleration toward the center and average orbital velocities of stars within any galaxy.

## Time; Relative to Light; Negative Direction

Our understanding of the universe is relative to light or light's constant speed; the time itself is relative to light; the constant velocity of light is 299,792,458 meters per second; so the moment light needs to travel 299,792,458 meters is 1 second. The time is the effect of the finite speed of light; If the velocity of light becomes infinite, then the time become zero or fail to exist.

In space-time the time is the emerging effect causing continuously by the velocity of light; the light carries observable information of continuous events or time; infact the light carries information of the past; if you observe objects or events at a distance of one light year then you are actually observing one year in the past; so the observable time is negative but the presently progressing time is positive; so the negative and positive of time is same or "The Time" because the negative and positive particle of light is same or the "Photon". Other hypothesis says big bang also occurred in the opposite direction and in there the time is negative. For a spherical universe, the time in the opposite direction to us is negative.

**The Dark Universe; Darkness of Physics**
The observable universe exists to us relative to time; so we observe everything in the past but what about the current present time? The things we observe now may have already destroyed. What is happening to the observed event in the current time? Maybe their current circumstances are totally different and even light from further distances can't completely reach to us. Our laws of physics are also relative to light or speed of light and so the time dimension is the finite speed of light; the wave properties of photons and particles is a mystery to us because of our understanding of time relative to light; every particle or everything exists as waves in the time dimension, which we still don't understand; the detectors exist in the spatial dimensions and so they can only observe the particle properties or spatial properties; observers of spatial dimensions only observe spatial properties of ordinary objects and affecting effects of the time dimension.

Nothing exists in the spatial dimensions without observation but everything exists within time dimension without observation; this is how everything of the unlimited universe is essentially existing everlastingly; ensuring events everywhere; everywithin; everywhen. These things are authentically beyond this book; there are so many mindblowing mysteries basically beyond presently progressing primitive clever cosmic characters.

## Interstellar Drive; Warp Drive

Spacetime is flexible and so it can be warped and that makes the warp drive possible; the warp drive warps forward spacetime continuously to travel vast distances faster than light; Enormous amount of positive and negative energies are required to cause continuous spacetime warping in order to travel faster than light speed. The understanding of negative mass is the key to produce negative energy.

## Timeless Teleportation; Intergalactic Drive; Wormhole

Spacetime is flexible and curvable; so wormhole tunnel can be created within spacetime; this tunnels will connect different regions of spacetime like faster than light travel tunnels, paths. Quantum teleportation can also create teleportation gates between galaxies; these gates can teleport objects from one galaxy to another one without depending on time; these teleportation gates require complete understanding of wave's configuration and conversion. Both of those technologies also requires enormous extreme energy or the energy of an entire star; multiple rapidly rounding round rings can remain stably near and around a star by gravitationally revolving it and also can constantly consume a star's entire energy; then that enormous energy can capably cause triumphant technologies that travel faster than light.

Ashraf A.

"Neverending Nineteen Nights Finally Finished; Forever
Fortunate Freedom's Fabulously Favorite Flags Flying
Flawlessly; Forwarding For Future For Forever."

162